ROUTLEDGE LIBRARY EDITIONS:
COLD WAR SECURITY STUDIES

Volume 41

SEAPOWER IN THE NUCLEAR AGE

SEAPOWER IN THE NUCLEAR AGE
The United States Navy and NATO 1949–80

JOEL J. SOKOLSKY

LONDON AND NEW YORK

First published in 1991 by Routledge

This edition first published in 2021
by Routledge
2 Park Square, Milton Park, Abingdon, Oxon OX14 4RN

and by Routledge
605 Third Avenue, New York, NY 10017

Routledge is an imprint of the Taylor & Francis Group, an informa business

© 1991 Joel J. Sokolsky

All rights reserved. No part of this book may be reprinted or reproduced or utilised in any form or by any electronic, mechanical, or other means, now known or hereafter invented, including photocopying and recording, or in any information storage or retrieval system, without permission in writing from the publishers.

Trademark notice: Product or corporate names may be trademarks or registered trademarks, and are used only for identification and explanation without intent to infringe.

British Library Cataloguing in Publication Data
A catalogue record for this book is available from the British Library

ISBN: 978-0-367-56630-2 (Set)
ISBN: 978-0-367-61147-7 (Volume 41) (hbk)

Publisher's Note
The publisher has gone to great lengths to ensure the quality of this reprint but points out that some imperfections in the original copies may be apparent.

Disclaimer
The publisher has made every effort to trace copyright holders and would welcome correspondence from those they have been unable to trace.

Seapower in the nuclear age

The United States Navy and NATO 1949–80

Joel J. Sokolsky

London and New York

First published 1991
by Routledge
11 New Fetter Lane, London EC4P 4EE

© 1991 Joel J. Sokolsky

Typeset in Times by Michael Mepham, Frome, Somerset
Printed and bound in Great Britain by
Biddles Ltd, Guildford and King's Lynn

All rights reserved. No part of this book may be reprinted or reproduced or utilized in any form or by any electronic, mechanical, or other means, now known or hereafter invented, including photocopying and recording, or in any information storage or retrieval system, without permission in writing from the publishers.

British Library Cataloguing in Publication Data
Sokolsky, Joel J. 1953–
   Seapower in the nuclear age: the United States Navy and NATO 1949–80.
   1. North Atlantic Treaty Organization. Role of United States
   I. Title
   355.031091821
   ISBN 0–415–00806–9

Contents

	Figures	viii
	Tables	ix
	Acknowledgements	x
1	Introduction	1
2	Establishing the NATO Maritime Alliance	7
3	The Cold War at sea: force and strategies	48
4	Soviet maritime forces and flexible response	76
5	Reinforcement sealift	125
6	The USN, NATO and the war at sea	177
	Notes	191
	Bibliography	203
	Index	216

Figures

2.1 Allied Command Atlantic July 1954	19
2.2 Allied Command Europe July 1954	31
2.3 Allied forces southern Europe 1961	32
2.4 Allied forces Mediterranean 1961	34
2.5 Allied Command Channel 1961	37
2.6 Commander-in-Chief allied forces northern Europe 1963	38
2.7 Allied Command Atlantic 1961	43
4.1 SACLANT in the 1970s	107
4.2 NATO military command structure elements in the Mediterranean and adjoining area	108
4.3 Allied Command Channel	112
5.1 NATO civilian shipping organization	152
5.2 Sequence of acquisition of US shipping	153
5.3 Acquisition of European NATO ships	154
5.4 Ship request work sheet	156
5.5 Ship request format (example)	157
5.6 Elements in German civilian control of shipping	164
5.7 Elements in German naval control of shipping	165
5.8 Sequence of NATO naval control of shipping	169

Tables

2.1 SACLANT's shipping requirements	23
4.1 Soviet Navy 1974	77
4.2 NATO maritime forces 1975	79
4.3 Major naval ship construction, Warsaw Pact and NATO 1955–74	80
5.1 Eastern hemisphere dry cargo merchant fleets	128
5.2 Particulars of ships with heavy derricks as at 31 December 1950	129
5.3 Representative sealift requirements	141
6.1 Comparison NATO and Warsaw Pact naval combatants 1983	182

Acknowledgements

In undertaking this study I have had the benefit of much support from numerous institutions and individuals. From 1978 to 1981, I was the recipient of a Canadian Department of National Defence Strategic Studies Scholarship which assisted me in pursuing graduate studies at the Johns Hopkins School of Advanced International Studies and Harvard University, where this study began as my doctoral thesis. I would like to thank my thesis advisor, Samuel P. Huntington for his guidance and patience. The Center of Canadian Studies, Paul H. Nitze School of Advanced International Studies of the Johns Hopkins University provided me with both employment and research support during the writing of my thesis. A NATO Research Fellowship allowed me to conduct research in Europe. A post-doctoral appointment at the Center for Foreign Policy Studies of Dalhousie University, Halifax, Nova Scotia allowed me to begin work on revising the thesis. Further research and final revisions were supported at the Royal Military College of Canada, Kingston, Ontario by a generous Arts Research Program grant. Secretarial and other assistance was also received from the Centre for International Relations of Queen's University in Kingston where Shirley Fraser, Bernice Gallagher and Marilyn Banting helped with typing and editing.

The bulk of the archival research was conducted at the United States Navy's Operational Archives at the Naval Historical Center, Washington Navy Yard, Washington D.C. The assistance of its staff was indispensable. I should also like to acknowledge the help of the Naval Historical Collection of US Naval War College in Newport, Rhode Island for directing me to the *Colbert Papers*.

Many serving and retired naval officers from allied navies agreed to be interviewed and several organizations arranged briefings for me. To those who gave graciously of their time and experience, I am greatly appreciative. In particular, I would like to thank Mr Roger Evans, formerly of the staff of the Supreme Allied Commander Atlantic who read early drafts of each chapter and made numerous suggestions.

Throughout my studies, I have been bolstered by the support of my family. My late parents, Leonard and Rose Sokolsky encouraged me to pursue higher education and were always there to help me in whatever way they could. My mother- and father-in-law, Mary and Clyde Hudnall have likewise provided encouragement. My children, Jared, Mark and Rachel, continue to be a source of immeasurable fascination and joy.

This work is dedicated to my wife Denise to whom I owe the greatest debt. During all the research and writing, her confidence and support has never wavered. Words alone could never capture my gratitude.

For any errors, omissions and unsound analysis in what follows, I alone am solely responsible.

Kingston, Ontario, Canada

Chapter 1
Introduction

The great naval debate of the 1980s concerned the United States Navy's (USN) 'maritime strategy'.[1] It was a broad concept for the conduct of global war – a war in which it was assumed Europe would still be the Soviet Union's prime target and objective. The goal of the strategy was to 'use maritime forces in combination with the efforts of our sister services and the forces of our allies to bring about war termination on favorable terms' (Watkins 1986: 4). This would be done by denying the Soviets the luxury of concentrating all their forces against Western Europe. The USN would horizontally escalate the conflict, engaging Soviet forces and those of their allies throughout the world.

In addition, rather than wait for the Soviet Navy to mount an offensive against the North Atlantic Treaty Organization's (NATO) maritime forces, the strategy called for forward operations in the Soviet Union's 'backyard'. These operations would include strategic anti-submarine warfare (ASW) against Soviet nuclear-powered ballistic missile submarines (SSBN). Not only would this reduce Soviet strategic nuclear forces and change the correlation of forces in the West's favor, it would help the Alliance at sea. In the absence of such pressure by the USN, it was argued, the Soviet Navy 'could concentrate major forces against supply lines we need for NATO reinforcement and other naval forces' (US Dept of Defense 1984: 9).

While its supporters saw it as being wholly consistent with, and indeed necessary for, the American commitment to the security of Western Europe, opponents saw it as a 'strategic misstep' that also went against the traditional postwar coalition and continentalist thrust in US defense policy and therefore undermined deterrence in Europe. Partisans of the 'maritime strategy' called forth the memory of Alfred Thayer Mahan in extolling the uniqueness and efficacy of seapower, forwardly and globally deployed, in the defense of Western Europe. Coalitionists, who often saw themselves as defenders of the Atlantic vision, warned of the strategic and political dangers of ignoring the continuing importance of Soviet land and airpower. The latter also charged that the 'maritime strategy' was a forerunner of increasing American unilateralism and estrangement from NATO.[2]

Often neglected in this debate was any appreciation of the history of NATO from a maritime perspective. In reality, the Alliance was a maritime coalition from its very beginnings during the Cold War. Seapower, both nuclear and conventional, both forward and high-seas, applied unilaterally by the USN and multilaterally in collaboration with the Royal Navy (RN) and other allied maritime forces, constituted a component of its overall deterrent posture. Indeed the USN's 'maritime strategy', promulgated with such fanfare in the mid-1980s, contained much the same thrust as NATO's Concept of Maritime Operations (CONMAROPS), which had been adopted by NATO in 1980. CONMAROPS stressed the importance of containing Warsaw Pact forces through forward operations, of defense in depth and of gaining and maintaining the initiative at sea (NATO 1981a: 19–20; Grove 1989: 3–5). But CONMAROPS itself did not represent a radical revision of the approach taken by the American and other NATO navies towards the defense of Europe. This study deals with the strategic, political and organizational considerations that informed the USN's role in NATO's maritime strategy in the period 1949 to the late 1970s.

The role of the USN for presence and crisis management, as well as its place in nuclear deterrence, has been extensively dealt with in the strategic literature from the Cold War on. However, it is only in the last decade that the problems of seapower in the NATO context have come to the fore. Before that the allied maritime posture was the 'forgotten front'.[3]

In some ways, this can be attributed to an underlying Mahanian influence. Since the Soviet Union did not possess a large modern navy in the early days of the nuclear age, and Western sea control was almost absolute, the Alliance would not be faced with the prospect of major sea battles. Thus the place of maritime forces in the overall NATO posture was not worthy of substantive discussions. It was only when the Soviets developed forces capable of challenging the West for control of the seas that attention was drawn to the allied maritime posture, or rather to its inadequacies.[4]

NATO became a maritime alliance precisely because it was in the first instance a coalition directed towards the maintenance of a balance of power in Europe. The main impetus for the creation of NATO came from the presence in Eastern Europe of large Soviet ground and air forces. In subsequent years, the Soviets added significant nuclear forces. From the Western perspective, these forces constituted a threat that could be met only by the aggregation of its own forces into a peacetime alliance, a latent war community, the principal aim of which was and is to 'protect the territorial inviolability of NATO territory' (Ørvik 1963: 6). The aggregate forces of the Alliance, especially the nuclear forces of the United States, provided a deterrent which not only dissuaded the Soviet Union from attacking NATO, but also undermined the credibility of threats of attack. This, in turn, negated the political influence which the USSR is believed to have sought over Western Europe.

NATO has relied upon a 'triad' of deterrent forces – strategic nuclear, theater-tactical nuclear[5] and conventional – to secure its primary objectives on the European mainland. Maritime forces constitute a part of all three elements of the NATO triad.

At the strategic nuclear level, the Alliance came to rely upon the US Navy's sea-based nuclear projection capabilities to supplement American long-range bombers and intercontinental ballistic missiles (ICBM). This entailed not only a general reliance upon all American sea-based nuclear forces, but a specific commitment. In the early 1950s this entailed aircraft carriers with nuclear delivery capabilities. By the 1960s, the US had begun to earmark a number of nuclear-powered ballistic missile submarines (SSBN) carrying submarine-launched ballistic missiles (SLBM) for Alliance use. To a lesser extent, NATO relied upon British and French SSBN forces.[6]

Maritime forces were also included in NATO's theater-tactical nuclear posture. Here the forces were dedicated both to projection and to securing use of the seas. American carriers continued to maintain a nuclear delivery capability after the deployment of the SSBNs, and, more recently, the US began to develop sea-launched cruise missiles (SLCM) which can be fired from submarines and surface ships (see Center for Defense Information 1981: 17). The SLCMs can be used in a tactical capacity against shore targets or enemy maritime forces at sea. Since the 1960s a number of the NATO navies have also deployed anti-submarine weapons which carry a nuclear warhead and which could be used against enemy submarines attempting to cut the sea lanes of NATO commerce and sink allied aircraft carriers.

In a war which quickly escalated to the theater or strategic nuclear level, the importance to NATO of using the seas would be limited, generally, to the requirement of being able to project nuclear force ashore. In the case of SSBNs, only the immediate firing area, which would be far out to sea, would have to be controlled, and then only for a short time period. Carriers would require a greater measure of sea control closer in for a relatively longer period of time.

There have been, however, expectations that NATO might have to sustain conventional war fighting forces on the ground, for an indeterminate length of time; this highlighted the Alliance's need to secure, deny and exploit the seas. Thus NATO has, since its beginnings, included maritime forces in its conventional deterrent posture. Indeed, most of the world's modern maritime forces belong to the NATO allies and most of these have been dedicated to supporting conventional deterrence on the European mainland. This applied especially to the US Navy. As a 1976 US Congressional Budget Office study of American naval force alternatives noted: 'A Navy strategy which presumes a prolonged conventional war in the Atlantic/Mediterranean region, generates requirements for large forces' (US CBO 1976: 29).

Allied maritime forces placed particular emphasis upon securing the seas for reinforcement and resupply. This meant development of large numbers of anti-submarine warfare ships, planes and attack submarines. It also entailed countermine forces to keep harbors open, and ground units specifically earmarked to secure coastal stretches bordering on strategic straits. In the Mediterranean, the US Sixth Fleet, although retaining a nuclear strike capability, had also been dedicated to providing air support for allied ground forces. Over the years, NATO developed a complex system for the acquisition of merchant shipping in the event of war for the transport of US reinforcements.

The Alliance created two of its three Major NATO Commanders (MNC), Supreme Allied Commander Atlantic (SACLANT) and Commander-in-Chief Channel (CINCHAN), specifically to engage in peacetime planning, and, if necessary, wartime control of the bulk of its collective maritime forces in the Atlantic Ocean and northern seas. Under SACEUR (Supreme Allied Command Europe) several subordinate commands were created which oversaw allied maritime forces, particularly those in the Mediterranean, including the US Sixth Fleet.

It was argued that during the Cold War the vast array of allied maritime forces which supported the NATO conventional posture reflected unrealistic expectations concerning the nature and duration of a conflict with the Warsaw Pact. Nuclear weapons would be used sooner rather than later, or allied ground and air forces would be unable to hold long enough for seapower to appreciably affect the course of the ground and air war. Such doubts highlighted the uncertainty that has always attended NATO strategy, an uncertainty that became even more pronounced with the growth of

Soviet capabilities, including the USSR's ability to deploy maritime forces capable of challenging NATO for control of the seas.

The uncertainty surrounding NATO's overall posture also emphasized the continuing relationship between seapower and the political and military situation ashore. As Christoph Bertram wrote in his introductory chapter to *New Strategic Factors in the North Atlantic*: '[I]t is the chances and likely outcomes of conflict on the landmass of Europe that is decisive', military considerations will 'determine the use of naval forces'. The relevance of maritime activity will be defined by the 'actual balance of forces in Europe and the balance of reserves and reinforcements outside the war theatre' (Bertram and Holst 1977: 17). It is true, therefore, that unless the Alliance maintained a conventional resistance for more than several weeks, maritime forces would not be able to influence the situation ashore in the event of a war. And, if it is a widely held peacetime perception that conventional deterrence lacks credibility, so too would maritime preparations tied to supporting conventional deterrence. On the other hand, maritime forces able to secure and exploit the seas for conveyance and projection could, under a given set of circumstances (i.e. extended strategic warning and prolonged mobilization) be an important factor in determining whether NATO would indeed be able to offer sustained conventional resistance and avoid having to make an early decision on the use of nuclear weapons. Moreover, given the politics of the Alliance, and what would undoubtedly be a time of great political confusion, the deployment of conventional maritime forces would necessarily have to take place during times of rising East–West tension, regardless of expectations as to the nature and length of an impending war.

In reality, the wartime efficacy of seapower in the NATO context could not be known until a war took place. In the meantime, judgements concerning the importance of maritime forces to the allied deterrent posture rested upon various calculations common to all other aspects of NATO's aggregated military forces. Partly out of the uncertainty which necessarily attends those calculations, NATO's political and military leaders elected to maintain a vast array of maritime forces, the bulk of which would only be truly effective in a conventional war. NATO's Cold War force structure was designed to meet the requirements of a long war. This made it equally applicable to short war and for purposes of political demonstration in crisis – short of war – situations. A force designed only for a short war or for limited political exigencies could not fight a long war. In such a war NATO's ability to secure and exploit the seas could be crucial, as would the ability of the Soviet maritime forces to deny or severely restrict allied control of the seas. The Alliance's conventional maritime forces represent another attempt by NATO to be prepared to meet as wide a range of conflict scenarios as possible and, by so doing, deter their occurring.

Together with the nuclear components of its collective maritime forces, those dedicated to conventional deterrence made NATO a maritime alliance comparable with the great maritime coalitions of the past. Like the coalition which finally defeated Napoleon and the alliances which twice defeated Germany, NATO developed its military forces in order to deter and, if necessary, wage war on the European mainland. Because of this overall strategic requirement, and not merely because its members depend to varying degrees upon seaborne trade, the Alliance maintained maritime forces whose tasks in war would be to secure, deny and exploit the seaward approaches to Europe. As in the other instances in which seapower was involved in a European conflict, the relative importance of NATO's maritime forces would depend upon the degree to which they could influence the situation ashore. In peacetime this entailed strengthening the allied deterrent posture; in the event of war this meant the extent to which allied maritime forces could support NATO land and air forces. It was also the case, that, as in the past, the situation ashore would determine the efficacy of seapower.

Whatever judgements are made concerning the strategic soundness of the allied maritime posture, the priority which should be attached to maritime force development, American allied maritime forces were part of NATO's overall posture since 1949. Given the overwhelming importance of the NATO and Warsaw Pact balance in the postwar international environment, it can be said that the incorporation of these forces into the Alliance's deterrent triad constituted the most significant and comprehensive example of seapower during the Cold War, and indeed, in the nuclear age.

Chapter 2
Establishing the NATO Maritime Alliance

SEAPOWER AND THE DEFENSE OF EUROPE: THE EARLY ASSESSMENTS

The early days of the nuclear age seemed to be at once a Mahanite's dream and his nightmare. On the one hand the United States and its allies held unquestioned 'command of the seas'; on the other, the advent of the atomic bomb appeared to make navies obsolete – superfluous – in any future war against the most likely enemy, the Soviet Union (Colletta 1980a: 480). It seemed at the time, and to several later analysts, that the 'whole hope' for the future of seapower lay in 'gaining a role in the strategic strike' (Denis 1971: 18). While the Western navies, especially the US Navy, were looking for such a role, the perceived nature of the political-strategic environment necessitated a much more complex approach to the future of seapower.

As the declassified documents now indicate, American naval planners in the immediate postwar years were extremely pessimistic about the United States' ability to defeat a Soviet ground assault in Europe despite the US atomic monopoly. This is a view naval leaders shared with the military high command. Those charged with the nation's security 'saw themselves confronted by a nearly unstoppable Russian war machine'. They expected, in the event of war, that they would initially be forced to abandon all of Western Europe, the Mediterranean and the Middle East. Strategic debate seemed to focus on the question of where it would be best to establish a toehold which could be used to 'neutralize and roll back' the Soviet onslaught on the ground (Rosenberg 1976: 61).

It is essential to note at the outset, therefore, that maritime planning in the early years of the nuclear age had as one of its basic assumptions the continuing importance of land power. The major threat to American security was the large conventional land armies of the USSR. Seapower had to assist the West in defeating these Russian forces. This recognition of the crucial role of land power in a future war appears in American maritime planning and thinking in the pre-NATO period.

In a memorandum to President Truman dated 14 January 1947, Admiral Forrest P. Sherman outlined what he believed would be the Navy's role in a future war with the Soviet Union.[1] Given the strength of US forces in Europe, and the condition of the Western European nations, any major assault by the Soviets would quickly overrun Germany, France, Belgium, Holland and Denmark. The United States would be forced to withdraw its forces to Spain, the UK and North Africa. There would then begin the mobilization and build-up on the periphery of Europe. All this in preparation for an eventual return.

The approach outlined by Sherman was incorporated into the US Navy's tentative operational planning in 1947. That July, Chief of Naval Operations (CNO) Chester Nimitz sent a message to all American commanders stressing that in the event of war the first task would be to secure the seas in order to evacuate US forces and available personnel from Western Europe, Italy and Korea. ASW (anti-submarine warfare) operations were to commence immediately so that the lines of communication across the Atlantic, and especially off the European coast, would be open. Bases were to be established in Iceland and the Azores for this purpose. In the Mediterranean, offensive operations would begin immediately.[2] Sherman had told Truman that a task force of sixteen carriers would be necessary to accomplish these essential sea control missions. The planes from the carriers would also be used to help slow down the movement of Soviet forces through Western Europe.

The most elaborate pre-NATO forecast of US naval tasks in the event of war was contained in the 1948 General Board study, *National Security and Naval Contributions Thereto* (US Navy 1948). The Report stressed that Western Europe held the balance of power between the US and the Soviet Union and that the main threat to Europe came from the large Soviet land armies behind the Iron Curtain. For the time being, the nations of Western Europe were too weak to hold the line against a Russian assault. Yet, every step must be taken to prevent the loss of Europe, for in a war with the Soviets the 'ultimate decision' would be reached there. 'Allied strategy,' therefore, 'must be based upon the greater mobility of our forces and take full advantage of technical superiority and exploit weaknesses in industry and logistics' in the Soviet posture. This meant immediate securing of the sea lanes, conducting ASW to protect shipping and establishing advance bases, all of which would allow the US to go to the offensive as quickly as possible. The offensive would include initiation of strategic nuclear bombing and, as soon as possible, the launching of a ground offensive once the Soviet position weakened (ibid.: 32, enc. D). The Board also argued that the Navy could play a crucial role in the first days of a war, 'more important than at any later time', not just in protecting sea lines of communication, but in using carrier airpower to help destroy the 'Russian power to advance'. With American ground air bases neutralized, 'carrier task forces may be the only striking power left' in the initial phase of a war (ibid.: 53, enc. D).

The Board did not suggest that carrier airpower would replace land-based airpower, particularly for strategic bombing purposes. In fact it stressed the importance of land-based airpower to ultimate victory. But the Report did warn that while strategic bombing would have tremendous influence on the outcome of a war with Russia, the 'capitulation of a nation who starts a war with full knowledge of the damage which [atomic] bombing can cause is not likely' (ibid.: 9, enc. D). The Navy's view was that the Soviets would eventually have to be pushed back on the ground by American and allied counter-attacks.

The ability of the United States and potential allies to 'take the war to Russia' would depend on the outcome of what was expected to be a 'vigorous' Soviet submarine offensive directed against shipping, particularly in the Atlantic (ibid.: 3, 11, enc. D). In his submission to the General Board Study, CINCLANTFLT (Commander-in-Chief Atlantic Fleet) had stressed that shipping 'in support of American and allied military campaigns as well as support of allied economies' would be necessary 'at the very outset of a war'. Convoying would follow the patterns of the Second World War with the number of ships equal to that used at the end of the war. He estimated that in a war, 81 ships per month would leave American eastern ports for northern Europe, Iceland and Greenland, with 249 per month leaving for the Mediterranean and southern Europe.[3] Plans developed by the Joint Chiefs of Staff (JCS) in the late 1940s had envisioned the movement of 22 divisions to the European periphery within a year after a Soviet attack. Planes and other supporting elements would also have to be transported.[4]

Given the tasks the Navy would have to perform in the event of war, the Soviet submarine fleet, as well as Russian naval aviation, constituted a significant challenge despite the unquestioned paramountcy of the United States in the quantity and quality of maritime forces. The prevailing view was that the battle for control of the Atlantic would not be a matter of two fleets meeting on the high seas and finishing the fight on the spot, with the US guaranteed victory. Rather, it would be a question of projecting American power on to the periphery of Europe in the face of formidable Soviet undersea forces.

Estimates of the numbers and capabilities of the Soviet submarine fleet varied. A 1949 study by the staff of the US Air Commander-in-Chief Atlantic (AIRLANT) put the number of Russian submarines capable of moving out any distance into the North Atlantic at 306. It also noted that the Soviets were building the German type XXI boat whose range, staying power and high speed exceeded their other submarines. AIRLANT reported 20 of these submarines ready for use in 1949.[5]

However, a report prepared by Admiral Low for the Chief of Naval Operations and released in April 1950 provided a lower estimate of Soviet submarine capabilities based on figures provided by naval intelligence. The *Low Report* put the total number of modern submarines (under 14 years) at 225, with less than half of these, 78, capable of 'ocean patrol' (i.e. endurance of 7,400 to 15,000 miles). Moreover, the report placed the number of 'high submerged speed' type XXIs available for use at no more than four, although it did note that countering this type exceeded US ASW capabilities (US Navy 1950).

While the *Low Report* thus reduced earlier estimates of the Soviet underwater capability and, indeed, noted that given the inferior design of Russian submarines, the Soviet force was less formidable than the German force in the Second World War, it warned that the USSR could still do serious damage in the North Atlantic in the event of war. 'The initial advantage inherent to the aggressor,' the *Report* concluded, 'coupled with our own weak escort forces, inadequate until after the Reserve fleet has been reactivated, would enable these submarines to inflict heavy losses upon our shipping in the early part of the war' (ibid.). Also of concern to sea lanes' security was the Soviet Naval Airforce, which was capable of laying mines and attacking convoys.

It was anticipated by American naval intelligence that even with their short range, Soviet submarines would be immediately employed at the outset of any hostilities on the ground in Europe. They would seek to cut off the movement of American military shipping through the approaches to the United Kingdom, in the Bay of Biscay and in the Western Mediterranean. In addition, 'some Soviet effort will be devoted to unprofitable areas for psychological purposes to tie down defense forces and to force convoying with attendant reduction in effectiveness of shipping'.[6] There would be an initial burst of submarine activity, lasting perhaps a month, followed by a lull while Soviet boats on patrol returned to bases.

It was believed that the Soviet submarine threat could be contained early in a war, and kept under control, provided sufficient capabilities were maintained. This included the ability to wage an aggressive anti-submarine campaign, employing carrier strikes, aerial minelaying and anti-submarine submarines. Also required would be barrier patrols, convoy escorts and ASW hunter-killer groups. A 1950 report by the Air Warfare Division of the Deputy Chief of Naval Operations (Air) concluded that, given the US ability to contain the Soviet submarine threat, enemy airpower rather than enemy submarines was the most serious threat confronting the Navy in terms of sea control (Rosenberg 1978: 261) in the Eastern Atlantic and Mediterranean. This did not, however, remove the need to maintain an effective ASW capability, nor the desirability of co-ordinating sea control measures with other Western nations.

In order to insure effective control of the North Atlantic in the event of war, the US Navy, in 1947 and 1948, began co-ordinating planning with the Royal Navy and the Royal Canadian Navy (RCN). Initial meetings aimed at 'closer standardization in the field of planning, operations and logistics'.[7] In April 1948 the three countries' naval chiefs met to co-ordinate planning in the event of war in Europe. Reports of the plans discussed indicate an assumption that all three navies would be involved in securing the lines of communication between North America and Europe at the outset of hostilities (US Dept of Defense 1979a: 280).

The war scenario pictured by the US Navy in 1949 was, of course, the same as that which took place in the European theater during the Second World War: the withdrawal from Europe, the build-up on the periphery, the convoying of material and the re-entry. Yet, the Navy's view was very much in line with the basic American War Plan of that time. Plan OFFTACKLE had been developed with the tacit agreement of Britain and Canada (ibid.: 289). Like its predecessors, OFFTACKLE envisioned a massive Soviet offensive in Western Europe and the Middle East, an attack which could not be contained, thus necessitating a withdrawal to the periphery. At the 1949 Brussels meeting of the NATO military planning group, the US representative told the Europeans that they would not expect American reinforcements until the later stages of a war. The version of OFFTACKLE then in place called for 41 US divisions and 63 tactical air groups to be sent to Europe for a re-entry attempt two years after the initiation of hostilities (ibid.: 294).

The allied counter-offensive would largely be shaped by the outcome of the strategic air war and the ASW campaigns and whether the Soviets invaded Iberia and the oil-producing areas of the Middle East. It was expected that 292 atomic bombs and 17,161 tons of conventional bombs would be used against the Soviets within the first three months. This strategic bombing would be primarily aimed at disrupting the Soviet war-making capacity and slowing down the advance of Soviet armies in Western Europe. Attacks on refineries were planned, but studies indicated that the Soviets would probably have enough oil to fight for a year and that the effect of the strategic bombing of refinery capacity would become evident only in a long war (US Dept of Defense 1979b: 164, 167). The JCS believed that the stockpile of nuclear weapons was still too small for SAC (Strategic Air Command) to strike one massive knockout blow. 'Therefore,' notes the JCS, 'conceived within the context of a protected conventional conflict, the experience of 1939–1945 was highly relevant' (ibid.: 165).

Part of the experiences of the 1939–45 war was the crucial role played by maritime forces in the North Atlantic in keeping the lines of communication open and the flow of supplies coming. Thus, of 'key' importance in the JCS view was the retention of 'those base areas ... and sea areas essential for offensive operations, including those required for launching and supporting the air offensive ... and to defend trans-Atlantic communications' (US Dept of Defense 1979a: 393).

To be sure, all planning at this time was based upon less than perfect knowledge of real Soviet capabilities and various estimates of allied capabilities in future years. Joint plans also reflected input from the three services and called for the full array of military measures such that the Army and the Navy as well as the Air Force would have vital roles in a war with the Soviet Union in Europe. Thus, to some extent the emphasis on protection of the sea lines of communication reflected the US Navy's efforts not to take second place to the Air Force – an effort that would continue.

Part of the Navy's concern was that it would be left out of strategic offensive roles for which its carrier task forces were suited. Thus in the JCS plan DROPSHOT, drawn up in 1949 for a war in 1957, the emphasis is on sea-based offensive operations against Soviet naval facilities as the best means to secure the seas. In this plan it was acknowledged that even with a projected 300–350 ocean-going submarines and 200–300 coastal submarines, Soviet forces 'would probably not be of sufficient strength to challenge openly allied control of the sea'. Nevertheless, this force would be capable of 'harassing attacks and of serious interference with sea LOC's in limited areas' and, unless they were destroyed, 'the existence of these forces would require uneconomical diversion of heavy units to convoy duty'.

DROPSHOT called for an immediate offensive against Soviet coastal facilities and naval yards in the Murmansk area and in the Black Sea in order to contain Soviet submarine forces as well as their naval aviation. Forces dedicated to this task would also be able to disrupt Soviet coastal shipping and effect a blockade of Soviet ports. A total of 40,000 mines were to be deployed at key areas during the first six months of a war. The carrier-based offensive would also prevent the Soviets from moving into Norway and perhaps taking key islands such as Spitsbergen and Iceland. As many as six carriers were to be deployed to the Barents–Norwegian Sea areas on D-Day and eight to the Mediterranean (including British contributions) for immediate strikes as part of the 'offensive operations to destroy enemy naval forces' (Brown 1978: 163, 204–5, 210–11).

While the USN was emphasizing offensive measures as the best way to secure the seas, there was little disagreement during NATO's first years that some steps would have to be taken to provide a measure of capability against Soviet maritime forces. The Medium Term Defence Plan which NATO's military committee accepted in 1949 as the Alliance's basic strategy contained reference to the need to maintain control of the North Atlantic. Phase One of the Plan called for the stabilization of an initial Soviet attack. Phase Two, for the initiation of major offensive operations. Phase Three, continuation of major offensives until capitulation. Phase Four, final achievement of allied war objectives. 'To support this defense, control of North Atlantic sea and air lines would have to be assured by local air defense of Portugal, Greenland, Iceland, the Azores and the Faeroes against seaborne raids' (ibid.: 307).

The first meeting of the North Atlantic Ocean Regional Planning Group took place in October 1949. Present were representatives of the Chiefs of Naval Staff of Belgium, Canada, Denmark, France, Iceland, the Netherlands, Norway, Portugal, the United Kingdom and the United States. The report of the proceedings and decisions forwarded to the JCS indicates a traditional view on the part of the participants towards security in the Atlantic.[8]

The group, which was to report to the Military Standing Committee, was charged with drawing up plans for the 'unified defense' of the North Atlantic. Five planning sub-groups were created: (1) Offensive Operations; (2) Atlantic Ocean Lines of Communication; (3) Sea Lines of Communication between the Atlantic Lines of Communication and Northwestern Europe; (4) Sea Lines of Communication between the Atlantic Lines of Communication and Western Europe and West Africa; (5) Defense of areas not covered by other regional planning groups, i.e. Portugal, Azores, Madeira, Iceland and the Faeroes.

The Offensive Operations sub-group included only the United States and Britain, with a permanent French representative. Its terms of reference were to: 'Prepare broad plans for offensive action against enemy armed forces and shipping, their bases and port facilities, including attack at source, amphibious and airborne operations and offensive mining'. Those groups charged with planning for the defense of sea lines of communication were to deal with the 'organization, control and protection of convoys and independent shipping against air, surface and submarine attack, including counter-offensive, anti-submarine operations'.

On 29 and 30 November 1949, representatives of Belgium, Canada, Denmark, France, Italy, the Netherlands, Norway, the UK and the US agreed to a pooling of merchant shipping in the event of war. These ships would then come under allied authority. In addition, steps were taken to produce what would later become a combined allied shipping control manual. Canada, the US and Britain agreed to combine the *Routing and Reporting Instructions* book of the US Navy with the Admiralty's *Naval Control of Shipping in War* into a single revised publication (US Navy 1950).

In the late fall of 1949, there was as yet no increase in the American standing forces in Europe. This followed directly upon the outbreak of the Korean War in June 1950, which increased the importance of security in the North Atlantic and thus the consequent need for more maritime forces there and the necessity of a unified allied naval command. In other words, the crucial element in the evolution of NATO's naval posture was not so much the acceptance of seapower as it was the recognition of the continuing importance of landpower, ground forces dedicated to holding the line in Europe as far to the east as possible. For if the Supreme Allied Commander Europe was to accomplish his mission of holding as far to the east as possible, 'his lines of communication through the North Atlantic and his right flank in the Mediterranean must be secure' (US Dept of Defense 1979b: 230).

It is essential to see the development of the Alliance's maritime posture in the larger political and strategic context. Politically, the Alliance, resting on the American guarantee of European security, was meant to assure the independence of the Western democracies. This could be accomplished by negating the threat posed by the Soviet Union through a policy of collective defense and deterrence. In 1950 the chief threat came from the large Soviet land armies in Eastern Europe. Strategic nuclear strike forces could not by themselves provide a sufficient deterrent against the threat since they would not be able to halt a Soviet advance through Western Europe. Even US Air Force Chief of Staff Hoyt Vandenberg admitted to a Senate committee that without delaying ground forces, the Red Army could overrun Western Europe despite great losses in strategic bombing (ibid.: 224).

American plans as of 1949, which accepted that any Soviet attack would mean the loss of most of Western Europe, were based upon calculations as to the forces then available to the US and the condition of the European allies. But there could be little doubt about the centrality of Europe in the global effort to contain the Soviets. Because of this understanding, it was further understood that the abandonment of Western Europe in the wake of a Soviet attack, however militarily necessary, would carry grave strategic, hence political, risks. NATO's defense plan was premised upon the belief that the outcome of the land war in Europe would determine the political settlement of any conflict with the Soviets. In February 1949, acting Chairman of the Joint Chiefs of Staff General Eisenhower had warned his fellow chiefs that the US must hold the line at the Rhine or a substantial bridgehead. If this could not be done, they must provide for a 'return at the earliest possible moment to Western Europe in order to prevent Communization of that area with long-term disastrous effects on US national interests' (US Dept of Defense 1979a: 288).

It was the US Navy which argued forcefully for a strategy which was designed to hold the line in Europe as far east as possible. Chief of Naval Operations Louis Denfeld contended that the strategy of accepting withdrawal from Europe with its reliance on atomic weapons to pave the way for an eventual re-entry was completely at variance with US foreign policy and national objectives. It 'abandoned Western Europe to the Soviets without a struggle, thus in effect handing over the manpower resources and industrial capacity of the region for the use of the Soviets'. Moreover, the CNO argued that ultimate victory would be foreclosed by the ceding of so much territory (ibid.: 299).

At the time he made these statements, Denfeld was waging an interservice struggle with the USAF in Congress. He wanted to prevent the cancellation of plans for a new carrier and the use of the funds to construct a new generation of long-range bombers, the B-36. Among the arguments used by him was that carriers would be more useful in supporting ground resistance to a Soviet attack on Europe (ibid.: 336).

Thus Denfeld's strategic concept, and that of other allied naval leaders, was indeed one which contained a greater role for naval power than that which was envisioned in plans that accepted the loss of Western Europe in the first phase of a war. In both views NATO would have to seek immediate sea control. But if an all-out effort was to be made to hold at the Rhine, thus placing more emphasis on effective defense in the first phase of a war, then the role of the allied navies would be much more significant and a wider range of naval capabilities would be required. Naval power would have to compose a larger share of the Alliance's deterrent forces.

Furthermore, to hold further east meant that the convoying of troops and equipment would have to begin almost immediately, even when large numbers of American troops were forward based. This in turn meant that NATO's available naval forces had to be augmented since there would be a shortened period of mobilization. More forces would also be required since it could be expected that even with the immediate commencement of search and destroy ASW operations and mining, some of the forces convoyed to Europe would fall prey to Soviet submarines. In view of this, there would also be a greater requirement for peacetime surveillance of the North Atlantic and co-ordination among the allied navies. Additional carrier forces would be employed off the European coasts for strikes against Soviet forces inland. These strikes would also begin immediately and would continue after security of the North Atlantic had been assured and the Soviet submarine force eliminated.

There can be little doubt that in making the best case possible for naval power in the defense of Europe, Denfeld and other senior naval officers were also looking toward increased budgetary allotments for their services. However, too much emphasis on the bitter and public debate between the Navy and Air Force can obscure several important considerations. The first is that, as Walter Mills argued at the time, the 'plain consequence' of the strategic concepts being put forth by the Navy was not to elevate seapower to a 'new fetish' in a nuclear age, but to 'put much greater emphasis on sea communications, seaborne weapons and seabased combined operations' as an essential element of a Western deterrent geared primarily to dissuade a Soviet land assault on Europe. The best way to prevent another general war was to maintain ground power in Europe, supported by air and seapower, sufficiently strong to insure that any Soviet attempt to overrun it would prove unprofitable (Mills 1951: 382–3).

Recognition by the Alliance of the importance of the North Atlantic in NATO's maritime and overall posture came in 1950 when the decision was made to place responsibility for planning in that region under a unified command. In December of that year the Defense Committee recommended that a Supreme Allied Commander Atlantic should be appointed as soon as possible after the selection of SACEUR (US Dept of Defense 1979b: 231).

Several of the European allies held joint naval exercises prior to the establishment of NATO's naval commands. Exercises VERITY, ACTIVITY and PROGRESS, involving British, Dutch, Norwegian, French and Danish forces were conducted in 1949, 1950 and 1951 respectively. They tested convoy protection, ASW, communications and tactical procedures

(US Dept of Defense 1952: 30). In 1951, the United Kingdom, France, Italy and the United States held joint exercise BEEHIVE I in the Mediterranean. They tested combined convoy escort, surface capabilities, ASW and carrier strikes using French and British airfields as mock targets.[9]

As part of the program to rebuild the military forces of European allies, by 1954 the United States was providing funds and training for the European navies in order to meet these requirements. The USN's share of the military assistance budget in 1951 was only $100,764,000 out of a total of $1.14 billion. This reflected the Truman Administration's emphasis on building up European land and airpower. Naval assistance went to all the European allies with England receiving the least and France the most – almost $33 million to the latter in 1951. The larger allies, France and Italy, sought funds for destroyers and carriers, while the smaller nations, Norway, Holland, Denmark and Belgium, looked towards acquiring smaller ships and minesweeping, gunboat patrols, and local ASW.[10]

While the need for greater co-ordination in planning had been recognized, and efforts to rebuild the European navies were underway, the appointment of SACLANT did not come until January 1952. The delay arose primarily because of British reluctance to have an American as Commander-in-Chief of Allied Forces in the North Atlantic. 'Should we have fallen so far into the walks of humanity', asked Churchill, from the opposition benches in 1951, that there was 'no British Admiral capable of discharging this function?' (US Dept of Defense 1979b: 231).

The politics of the Anglo-American dispute will not be recounted here (see Poole n.d.). Suffice it to note that Churchill, upon replacing Attlee as Prime Minister, steadfastly refused to accept an American SACLANT until he was persuaded to do so by President Truman and Dean Acheson in January 1952, but only after he had reminded the President and Acheson that 'For centuries England had held the seas against every tyrant.' Could not America in the 'plenitude' of its power, bearing as it did the 'awful burden of atomic command and responsibility for the final word of peace or war ... make room for Britain to play her historic role upon that Western sea whose floor is white with the bones of Englishmen' (Acheson 1969: 602).

Eloquence was not sufficient to preserve SACLANT for the Admiralty. In compensation, the US did agree to the creation of a Channel Command with a British Commander-in-Chief. In addition, in 1953 Admiral Lord Louis Mountbatten of Burma was appointed Allied Commander-in-Chief Mediterranean. But the Alliance's principal weapon there – the US Sixth Fleet – remained under American control with its Commander, Admiral Carney, subordinate only to SACEUR. For the British, the establishment of SACLANT confirmed that the historic role of defending the 'Western Sea' (and all other seas) in the nuclear age had passed to the United States.

The US sought command of the Alliance's naval forces in the Atlantic because it believed that only a centralized command would be effective in supporting SACEUR. Given the magnitude of this task, in particular the threat posed by Soviet submarines, and the fact that America would supply 75 per cent of allied naval forces, it was imperative that SACLANT occupy the same command level as SACEUR, and that he be a US naval officer (US Dept of Defense 1979b: 280). The US had rejected the British proposal of dividing command of the Atlantic with a British admiral as commander in the East. There could be no separate command between the US area and SACEUR. Important as well was the demand of the Americans that the division of responsibilities between SACEUR and SACLANT be handled by later agreement between the two US commanders. The US was concerned, for example, that SACEUR would be able to control naval movements along his northern flank through the North Sea.[11]

When established, the Atlantic Command extended 'from the North Pole to the Tropic of Cancer and from the coastal waters of North America to those of Europe and Africa' with the exception of the English Channel and the waters around the British Isles. There was a division of command beneath the Supreme command level. The Eastern Atlantic was placed jointly under a British Naval Commander-in-Chief and a British Air Commander-in-Chief. SACLANT himself served as Commander of the Western Atlantic. Under the Commander Western Atlantic, there was a Canadian sub-area under the joint command of an RCN Admiral and a Royal Canadian Air Force (RCAF) Air Commander.

Aside from these area divisions, all subordinate to SACLANT, there was a separate subordinate operational command – the Striking Fleet Atlantic – a force of heavy surface ships, aircraft carriers and necessary supporting units. Its role in time of war 'would be to undertake offensive and support operations rather than the direct defense of the Atlantic trade routes'. The fleet would be available to furnish support to other NATO supreme commanders besides SACLANT (NATO 1955: 75) (see Figure 2.1).

SACLANT headquarters was organized into several divisions: Personnel and Administration; Intelligence; Plans, Policy and Operations; Logistics; Communications; Budget and Finance; and Public Information. Direct liaison was maintained with SACEUR through the SACLANT representative in Europe (ibid.).

The first SACLANT was Admiral Lynde McCormick, USN. It was

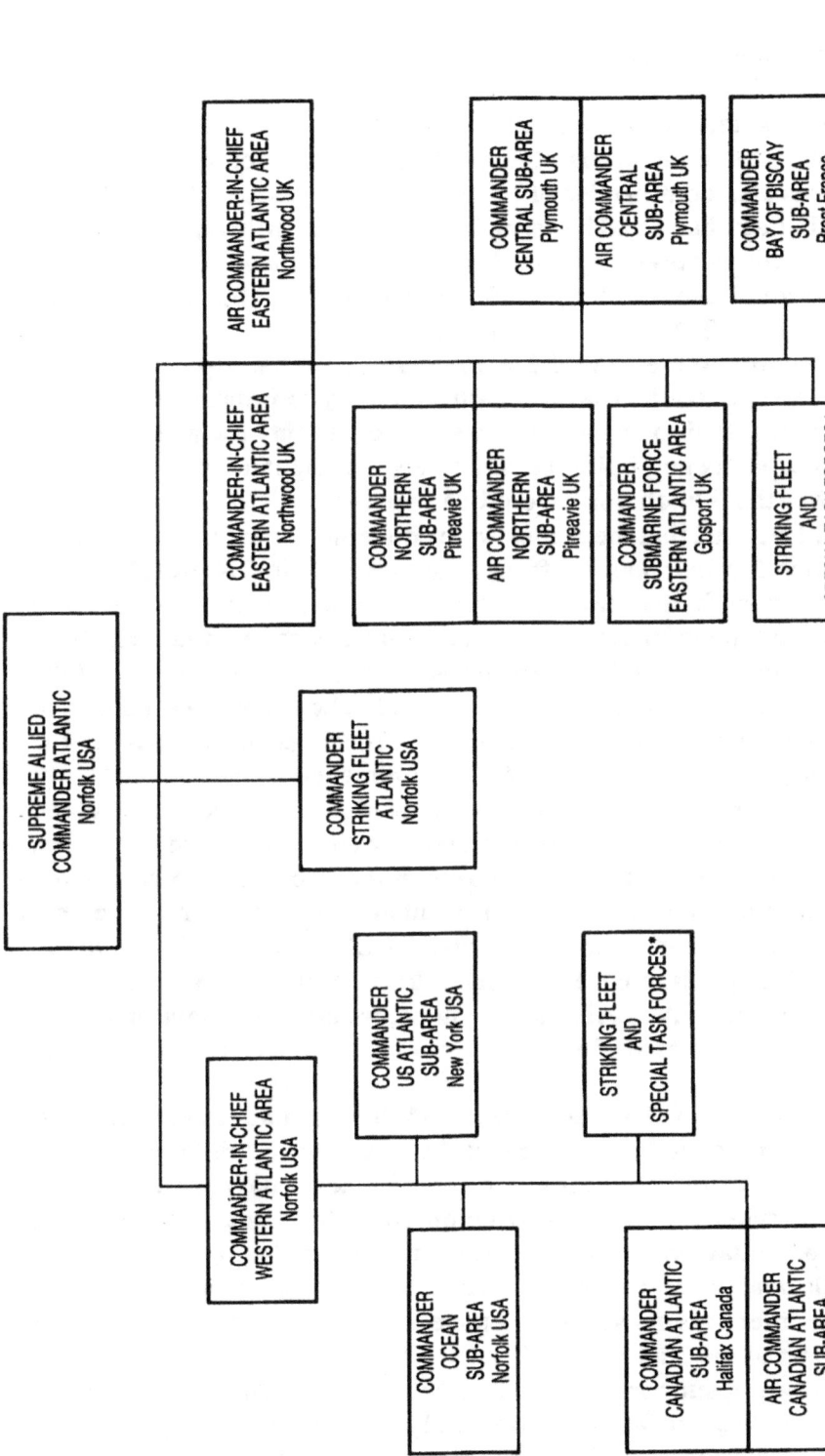

*When assigned

Figure 2.1 **Allied Command Atlantic July 1954**

said at the time that he was an Admiral without a fleet since allied naval forces would come under his command only in time of war. However, since he was also commander of all US forces in the Atlantic, including the Atlantic fleet, he would have, at the outset of hostilities, command of most of his striking power. It was important, though, that he obtain pledges of contributions from the allied navies. Shortly after taking up his post he traveled to all NATO capitals seeking such support.

In his report[12] to the NATO Standing Group on his trip, McCormick stated that he wanted the European allies initially to improve their capabilities to secure their own national waters in the event of war. The Dutch, for example, had begun building minesweepers (eighteen paid for by the US) in order that Dutch ports would be kept open (US Dept of Defense 1952: 44). He noted that the Danes were concerned about the availability of sea support for their national forces and those of CINCNORTH, should the Soviets attack in that direction. The French were anxious about their lines of communication off their whole Atlantic coast, particularly those to French Morocco. In addition, the French objected to the creation of the Channel Command under a British Admiral. They made the point that they 'do not desire convoys bound for French ports to pass to the operational control of other than NATO commanders'.[13]

While McCormick did receive varying pledges of support from allied governments it seems evident from his remarks that for the smaller allies, the NATO naval problem was primarily an American one to be worked out with the other two relatively significant naval contributors at the time – Canada and Britain. Indeed, the first SACLANT told the Standing Group that his appointment had been welcomed in Europe with a sense of relief: 'Now that SACLANT is appointed, we no longer have any naval worries, he will take care of everything for us ... we need not do anything now.'[14]

In his report[15] to the Permanent National Representatives in May 1952, McCormick drew attention to the fact that many were thinking solely in 'mathematical terms' with regard to the Alliance's naval needs. It was wrong to calculate those needs by comparing NATO naval forces with those of potential enemies. Requirements would be based upon the 'magnitude and diversity of the maritime task we must successfully perform in order to bring victory. Some of our most essential tasks would be in support of the land battle.'[16]

The lukewarm support that SACLANT received from some European nations was not surprising. They were only then just beginning to rebuild their navies. Moreover, the US Navy's analysis indicated that for at least the next few years the smaller European powers could not be expected to contribute to naval defense significantly beyond their coastal waters. In July 1952, the Director of the General Planning Group of the Office of the Chief of Naval Operations issued a report to the Deputy CNO on NATO naval readiness.[17] It was a forecast of available forces for a D-day sometime in 1953. In the realm of major combatants (carriers, battleships, destroyers, submarines), 80 per cent of the NATO effort would be exercised by the US alone, and more than 90 per cent by the US, Canada and Britain. The only available amphibious forces would be the two American Marine divisions assigned to CINCLANT. 'It follows, therefore,' the Director concluded, 'that the status of reserves for the few major combatants in the other eight NATO navies is relatively unimportant.' However, in terms of coastal escort and coastal and inshore minesweeping, 'the contribution of the eight is much more significant than of the three larger navies'. He believed that it was less important that the Dutch submarines be provided with one or more refills of torpedoes 'than that the Dutch mine and patrol craft be able to continue their assigned task in phase with war plans'.[18]

In the fall of 1952, NATO held its first major naval exercise, MAINBRACE. Conducted in the eastern Atlantic under the joint command of SACLANT and SACEUR, the exercise involved 160 allied ships of all types, including 9 carriers and 36 destroyers of the US Navy as well as Canadian destroyers and the RCN carrier MAGNIFICENT. The purpose of the exercise was to test SACLANT's ability to 'provide northern flank support for a European land battle'.[19] The assumption was that Germany had been overrun and that Soviet forces were moving into Denmark and northern Norway. Thus, in addition to its practical importance in terms of affording the opportunity to test capabilities and allied co-operation, this first exercise had an important political purpose. It was partially conceived to 'reassure the Scandinavian signatories (Norway and Denmark) that their countries could be defended in the event of war' (Kealy and Russell 1967: 85).

During his first year, McCormick also conducted several smaller exercises. GANNET and EMIGRANT, involving the Americans, Canadians, British and Portuguese, tested the ability of the NATO navies to control and protect shipping in the Atlantic against limited surface and submarine attacks. In February 1953, exercise BUFFET brought American, Canadian and Dutch forces together to test 'co-ordination between air and surface forces and between maritime patrol aircraft and submarine hunter-killer groups'. McCormick reported to the US CNO that 'operations of the hunter-killer groups were excellent; otherwise results were disappointing primarily because of poor equipment and lack of training of other national forces. Valuable training was received, however, in co-ordinating forces of different nationalities in working together as an ASW team.'[20]

In order to provide a co-ordinated approach to allied shipping in the event of war, NATO set up the Planning Board for Ocean Shipping. In addition, in each country in Western Europe, Ship Destination Rooms were set up. These organizations would decide at which destination ports convoys would dock before they sailed, and the diversion of ships in convoy to other ports if the original destination port became incapacitated. They would also be responsible for insuring links between the ports and the inland transportation system (NATO 1955: 145–6). In 1952, the NATO Council published its first *Allied Control of Shipping Manual*. On the basis of the manual, McCormick conducted two paper exercises to test the US Navy's Naval Control of Shipping Organizations' (NCSOG) ability to implement allied shipping plans. He noted that the 'impact of expanding responsibilities under NATO has increased the workload of Naval Control of Shipping Organizations considerably ... operations planning, training preparation and execution of exercises ... have continued at a high level for both national and NATO organizations'. As a further indication of the importance attributed to shipping, CINCLANT had begun a two-week training course for prospective convoy commanders and their staffs. By June 1953, 37 commanders and 50 staff officers had gone through the course.[21]

The emphasis on convoy protection in the exercises, planning and training begun under McCormick indicates that the Alliance anticipated immediate convoying upon the outbreak of a war in Europe. Further confirmation of this is found in the records of the American Joint Chiefs of Staff. In August 1952, McCormick requested that the JCS provide him with information concerning 'special' American convoys which planned to sail in SACLANT's area within ninety days of the commencement of hostilities.[22] The JCS reply[23] confirms that the commitment of American troops to forward defense in Europe had increased the need to begin immediately resupply and reinforcement. A military convoy of fifteen ships 'intended specifically for the initial support of US Army forces in Europe as contained in current plans' was to sail at D+15 days with disembarkation at Bordeaux. Preparations had been made to divert this convoy to a number of other ports, including Le Havre, Cherbourg, Marseilles, Barcelona, Casablanca and Oran, depending upon the immediate course of the land battle in Europe. Other special convoys were to leave the US every fifteen days from D+30 to D+60–5.

However, CINCLANT would not have 'sufficient escort forces avail-

Table 2.1 SACLANT's shipping requirements

Requirement	Shipping must be ready on	Sail with convoy departing on date shown
D+15	D+13	New York to UK – D+13
D+30	D+29	New York to UK – D+29
D+45	D+45	New York to UK – D+45 or Norfolk to Gibraltar – D+44
D+60	D+60	New York to UK – D+61 or
D+65	D+61	Norfolk to Gibraltar – D+60

able to permit the scheduling of special convoys for this particular purpose'. Therefore, his plans were changed so that shipping specifically intended for the support of the US Army 'will normally have to join regularly scheduled convoys'. Table 2.1 (reproduced from JCS files) shows the integration of shipping requirements as forwarded to SACLANT on 25 August 1952.[24]

SACLANT's shipping schedule was formulated on the basis of his Emergency Defense Plan developed in 1952 – EDP 1–52. Although the plan, and its subsequent revisions, remains classified, a number of declassified documents which make direct reference to it afford a good outline of NATO's overall maritime strategy at this time.

EDP 1–52 provided for the initial deployment of allied naval forces and detailed the tasks they were to perform for the period from D-day to D+3 months. It also provided a general plan for operations in the first year of a war and the estimated force deployments from D-day to D+6 months. The stated purpose of EDP 1–52 was to 'provide strategic guidance and the concept of operations within Allied Command Atlantic'. It was to establish doctrine, command structure and procedures, and to indicate the prospective assignment of earmarked forces for planning purposes so that operational commanders 'may prepare operational plans for execution should hostilities commence in the near future'.[25]

The plan envisioned that SACLANT would simultaneously have to seek to secure the North Atlantic lines of communication, provide convoy protection and use his striking fleet to support SACEUR in Europe. Sea control and ASW operations would be centralized. All contributing navies would be aware of the location to which to send their earmarked forces in the event of war. American and Canadian escort forces, including some carriers, would not be sent immediately into the open sea. Rather, these ships would move to selected North American ports in order to 'support early heavy convoys' of American reinforcements for Europe.[26]

In his emergency defense plan, SACEUR had called for close support from SACLANT to secure his lines of communication across the Atlantic and provide support for his ground forces.[27] SACEUR had also stressed the 'indispensability' of heavy carrier forces and 'the necessity that the support of these naval forces be co-ordinated into his overall operational plans for the area'. In January 1951, Eisenhower, in briefing President Truman and the Cabinet on his strategic concept for the defense of Europe, had pointed out the importance of carrier-based airpower in the Mediterranean and the North Sea. These forces would be used to hit the Soviets from the flanks if they moved ahead against allied ground power in the center (Rosenberg 1978: 265). In his March 1951 letter appointing Eisenhower SACEUR, Truman affirmed that the Sixth Fleet would be under Eisenhower's command 'to the extent necessary for the accomplishment of [his] mission' (Dur 1976: 37).

During the subsequent Anglo-American disagreement over Mediterranean commands, Eisenhower telegrammed CNO Sherman, stating that whatever the outcome of the command controversy: 'The US Striking Fleet [Sixth Fleet], in the Mediterranean shall be allocated to me for the defense of my right flank.... This fleet represents a great potential for application wherever needed in furtherance of the allied effort.' The fleet would also serve in the development of allied military capabilities and potential on the southern flank.[28]

SACEUR was especially concerned that he be guaranteed not only sufficient naval forces, but that the European theater be given priority call on American naval assets in the event of a general Soviet assault in Europe and Asia. Admiral Sherman assured Eisenhower in May 1951 that 'all our plans are predicated on maintaining essential provisions for Pacific and sending to the Mediterranean all large carriers'.[29]

Thus, with the strong support of General Eisenhower, the US Navy, which was then looking to assure itself a secure place in the new nuclear strategic environment, 'was virtually assured that it would play a major role in organizing NATO defenses in the Mediterranean' (Dur 1976: 37). The Sixth Fleet's primary mission changed from one of defending the sea lines of communication to one of, initially, conventional air support for allied armies. It would attempt to impede the movement of Soviet forces, provide direct tactical air support over the battlefield and utilize its amphibious marine component 'to seize an advance base or beachhead to prepare for the arrival of a division size amphibious task force from the United States' (ibid.: 39–40). At the Lisbon meeting in 1952, the US formally agreed to earmark two carriers, and their supporting elements, for CINCSOUTH upon receipt of a mobilization order from SACEUR.

At that time the US Navy had 2 carriers, 3 cruisers, 2 destroyer squadrons, along with supporting elements, in the eastern Atlantic and Mediterranean. In addition there were 4 carriers in the Atlantic with one more on ready-reserve. Another 6 carriers, plus one ready-reserve, were in the Far East and Pacific. According to a June 1951 memo to the CNO, in the event of a Soviet attack in Europe and the Middle East, 3 carriers would move from the Atlantic to the Eastern Atlantic and Mediterranean, and 2 would begin transit to the European theater from the Pacific. By D+30 days there would be 7 carriers assigned to the European theater with 2 held back in the Western Atlantic. Thereafter, as more carriers became available through mobilization, they too would be sent to European waters. By D+6 months the disposition of American carriers would be as follows:

Eastern Atlantic and Mediterranean	8
Atlantic Home	4
Western Pacific	4
Pacific Home	2

The 4 carriers in the Atlantic (Home) were to provide training and to 'constitute the back-up to insure that eight carriers can be maintained on the line in the Mediterranean–Europe area'. It was expected that 4 more carriers would be available by D+18 months. As they became operational they were to be assigned to SACLANT, resulting in the following disposition:[30]

Eastern Atlantic and Mediterranean	8
Atlantic	4
Atlantic Home	4
Western Pacific	4
Pacific Home	2

In September 1951, the JCS had expanded the role of the Navy to include provision of forces 'essential to the conduct of operations by Supreme Allied Commander Europe.... For example the Sixth Fleet, now in the Mediterranean, will provide naval support to SACEUR in the accomplishment of his missions' (Rosenberg 1978: 265).

The Joint Strategic Plans Committee (JSPC) of the JCS found SACLANT's plans for the provision of carriers to support SACEUR to be sound, and recommended that EDP 1–52 specifically provide for co-ordinated planning between the subordinate commanders of SACEUR and SACLANT.

However, when the plan was submitted to the Joint Chiefs for final approval there was disagreement between the Air Force and the Navy as to the scope of SACLANT's support for SACEUR, in particular the use of the striking fleet for strategic bombing of ports and Soviet naval facilities, including the use of atomic weapons.

In 1949, after the Harmon Report had demonstrated that strikes on industrial centers would not prevent the Soviet Union from taking Western Europe, and after the NATO Treaty committed the US to defend Western Europe, SAC was tasked with holding back a Soviet advance with nuclear weapons. However, 'plans for carrying out this mission were three years in the making'. Before 1952, the JCS did not feel they could allocate atomic weapons or units specifically for tactical use. This was because of the difficulty of hitting troop concentrations and the scarcity of atomic weapons. By 1952 there had been an increase in weapons production and the development of small atomic bombs which could be delivered by general-purpose fighters and day-attack planes like the Navy's F2H–2B Banshee and ADa–4B Skyraider. In February 1952, the Joint Chiefs told Eisenhower that atomic weapons had tentatively been allocated for tactical use in the defense of Western Europe and that Navy as well as Air Force aircraft could be considered as prospective delivery vehicles (ibid.: 266).

In his memorandum on EDP 1–52, the USAF Chief of Staff said he could support the idea of offensive carrier action against land forces only as a 'temporary expedient capable of providing a degree of support to SACEUR in the absence of sufficient land-based tactical airpower'.[31] Airpower from the carriers could be useful in 'retarding early Soviet movement' across Europe. However, the Air Force believed that SACLANT had to prioritize his tasks. The first task was securing the sea lines of communication. Since on D-day 75 per cent of the Soviet submarine force would be at sea, and at D+90 days one-third would still be at sea with the remainder *en route* and in port, the early destruction of yards and docks would not be likely to yield major returns for the first 180 days.

There was also a problem of targeting for nuclear strikes originating from US carriers assigned to SACLANT. The Air Force pointed out that with the exception of some air bases, the targets contained in EDP 1–52 were generally classified as strategic and 'considered for attack by the US Strategic Air Command'. Attack priorities for such targets should be considered by the Joint Chiefs 'in relation to the overall initial atomic effort'. Also to be considered was the fact that certain NATO land commanders were authorized to mark retardation targets for destruction with atomic weapons. Thus, the USAF Chief of Staff concluded, 'until such time as CINCLANT's atomic annex had been co-ordinated with appropriate US commanders, and approved by the Joint Chiefs of Staff, atomic targets ... proposed for destruction by SACLANT cannot properly be incorporated into the SACLANT plan'.[32]

In response, the US CNO argued that the carrier strike forces, as requested and outlined by EDP 1–52, were 'indispensable' to the accomplishment of SACLANT's missions.[33] They enable him directly to assist SACEUR. There could be no pre-hostilities prioritization of tasks for SACLANT:

> All of the elements set forth in SACLANT's mission are essential and each of them must be accomplished. If the convoys do not arrive in Europe, SACEUR's forces are doomed. If the Soviets outflank SACEUR to the North, SACEUR is in grave circumstances. If SACEUR's forces do not receive direct naval support, they may not be able to resist Soviet advances. Not only are these elements all essential, the successful accomplishment of one will assist in the accomplishment of the other. Enemy action, and the effect of that action on our allied forces, that is, the circumstances then prevailing, will determine where the emphasis should be placed at any particular time. It is certain that relative emphasis on the various elements of SACLANT's mission will vary as the battle for Europe progresses and NATO forces must be capable of shifting emphasis as the need arises.[34]

It would appear that EDP 1–52 was approved by the JCS, with the Air Force's reservation that the nuclear role for SACLANT await further planning. The plan itself indicates the extent to which SACLANT sought to provide an allied maritime posture capable of meeting any contingency brought on by a Soviet attack on Western Europe. If the attack was such that SACEUR was able to hold, then he had to be supported from carriers and resupplied through convoys. If SACEUR had to fall back, with the need for evacuation, the naval forces would have to cover his retreat and provide ships for evacuation. And, of course, SACLANT argued that the decision to hold or retreat would depend on how much naval support SACEUR could receive, including atomic strikes against selected strategic targets. It should be noted as well that at this time the US Navy believed that the Soviet submarine threat could best be met by attacking bases, if necessary with nuclear weapons.

NATO naval exercises in the years 1953 to 1955 reflect the broad view SACLANT held of the role of naval power in the Alliance's deterrent posture. From 16 September to 4 October 1953, over 300 allied ships took part in war game exercise MARINER. Submarines and some surface ships acted as the attacking forces trying to interdict convoys and prevent offensive operations. The exercise was jointly sponsored by SACLANT, Channel Command and SACEUR. As McCormick reported to US CNO Robert Carney, 'there was no strategic concept other than Blue was fighting Orange'.[35] The exercise was meant to test a wide range of allied naval capabilities: command relationships, communications, logistics, coastal and ocean operations, ASW, control of shipping, offensive operations from the striking fleet, mine warfare and intelligence. Actual convoys were formed in the Western Atlantic and run to the northern flank. The striking fleet ran several mock nuclear bombing runs on simulated targets in the UK. It was presumed that nuclear weapons would be available to both sides, and tests were made on defense against atomic bomb attack on the fleet.[36]

The exercise indicated to SACLANT that his command set-up was capable of meeting an emergency in the North Atlantic. Ships and aircraft of many navies worked well together under adverse weather conditions. Tests on hunter-killer techniques went well and allied submarines in the attacking force scored several kills. However, a number of problems were revealed. In the field of communications, many allies had not converted to UHF frequencies, hampering co-ordination. Another major problem was logistical support.

Logistical support had been one of McCormick's earliest concerns, and would remain one of the principal weaknesses of allied maritime strategy. Since SACLANT was composed of separate national contributions, logistical support for these ships was not directly under either a NATO area or operational commander. In October 1952, the American Joint Chiefs responded to McCormick's request for mobile logistical support by allocating three mobile support groups to SACEUR and SACLANT, to be made available immediately upon commencement of hostilities.[37] However, MARINER had revealed continuing problems with 'co-ordination and control of logistical support in a manner that would ensure responsiveness to combat operational commanders'.[38] A conference of all NATO naval commanders, held in November 1954, failed to resolve many of these difficulties.[39]

In addition to large-scale exercises, SACLANT conducted several smaller ones to test specific capabilities: NEW BROOM saw Canadian and American ships trying out advanced ASW equipment; TRADE WIND III tested allied control of shipping to selected destinations in Europe; NEW BROOM II was held in the Canada sub-area (September

1954) under Canadian naval and air commanders with US and Canadian forces participating in ASW exercises with an emphasis on convoy escort and air and sea co-ordination. A similar exercise was conducted in that area in May 1955.[40]

In April 1954, SACLANT met with representatives of the US CNO, the US Joint Military Transport Committee (of the JCS), the British Admiralty and the Canadian Naval Command. At this meeting, the allied convoy schedule was modified so as to accommodate US military lift requirements during the period D-day to D+30 days.[41] In 1955, exercise LIFE LINE, a naval control of shipping and logistics exercise, tested NATO command relationships and procedures for the protection of shipping. This exercise was jointly sponsored by SACLANT, SACEUR and allied Commander-in-Chief Channel and Southern North Sea.[42]

THE MEDITERRANEAN AND THE ENGLISH CHANNEL

In the Mediterranean[43] a somewhat awkward compromise was reached on command arrangements which sought to reconcile the fact that the US possessed the greatest naval striking power in the region, the Sixth Fleet, and British sensitivities based upon their continuing interests in the Mediterranean apart from NATO, as well as the need to compensate England for not being given the Atlantic command.

Originally, Eisenhower, upon assuming the position of SACEUR, had established a subordinate position, Commander-in-Chief Allied Forces South (CINCSOUTH). This post also went to an American, Admiral Robert Carney, who was then US Commander-in-Chief Eastern Atlantic and Mediterranean (CINCNELM). Carney created three subordinate commands to CINCSOUTH to direct the air, land and sea components of the allied forces earmarked for his command, with Carney himself holding the post of COMNAVSOUTH (Commander Allied Naval Forces South). The purpose of COMNAVSOUTH was to provide naval support for France and Italy, and its earmarked forces included the French Mediterranean fleet, the Italian Navy and the US Sixth Fleet. It did not include British naval forces in the Mediterranean, mainly because these forces were still primarily dedicated to protecting British interests in the eastern part of the sea – then outside the NATO sphere, although British and NATO forces held combined exercises at the time.

With the admission of Turkey and Greece into the Alliance in 1952, the maritime command structure in the Mediterranean was altered. The Greek and Turkish air and land forces came under CINCSOUTH, but the naval units were earmarked for a new NATO command, Allied Forces

Mediterranean (AFMED), which was created in December 1952 with the British Commander-in-Chief Mediterranean Fleet as CINCAFMED, a position subordinate to SACEUR and on a par with the other three regional commands. With the creation of AFMED, NAVSOUTH was abolished. The naval forces from France, Italy, Greece and Turkey were to be earmarked for the new command.

Excluded, however, was the US Sixth Fleet which remained under CINCSOUTH as STRIKFORSOUTH. Its task was directly to support the land and air campaigns in southern Europe with amphibious forces and carrier-based airpower, including tactical and strategic nuclear strikes. As Lord Mountbatten, NATO's first CINCAFMED, stated at the time: 'The carrier task force (6th Fleet) used in this mobile and flexible way is an extension of the strategic air force.'[44] In his 1951 report CINCNELM noted that the Sixth Fleet had altered its primary mission from protection of the Mediterranean sea lines of communication (SLOC), to 'projection of carrier air and amphibious power in support of SACEUR and his mission to defend Western Europe' (Dur 1976: 33). The protection of the Mediterranean SLOC was the responsibility of AFMED: 'It is not part of the duties of the striking force to support the war at sea' (Mountbatten 1955: 173).

There appear to have been two reasons why the projection role was separated from the sea control role and placed under separate Commanders-in-Chief. In the first place, given the poor state of allied ground and air units in the southern region, it was necessary to bring the Sixth Fleet directly under the control and in support of the regional commander who would be responsible for the ground war. This was especially the case with regard to Greece and Turkey.

The second and more important reason was that the nuclear capability of the Sixth Fleet had to remain under exclusive American command according to US law. While the fleet was earmarked for CINCSOUTH, in support of SACEUR's tasks in Europe, its nuclear strike forces belonged to CINCNELM as a major command directly subordinate to the US Joint Chiefs of Staff and, ultimately, the President of the United States. This was similar to the sea-based strike forces assigned to CINCLANT/SACLANT. Thus the decision to employ nuclear weapons could be made by the United States without recourse to the North Atlantic Council which was the supreme authority over the actions of all NATO commanders. In a strict sense, to have placed the Sixth Fleet under CINCAFMED would have placed this multinational decision-making body between the fleet and Washington. Leaving the fleet under CINCNELM put its nuclear forces on the same standing as the long-range

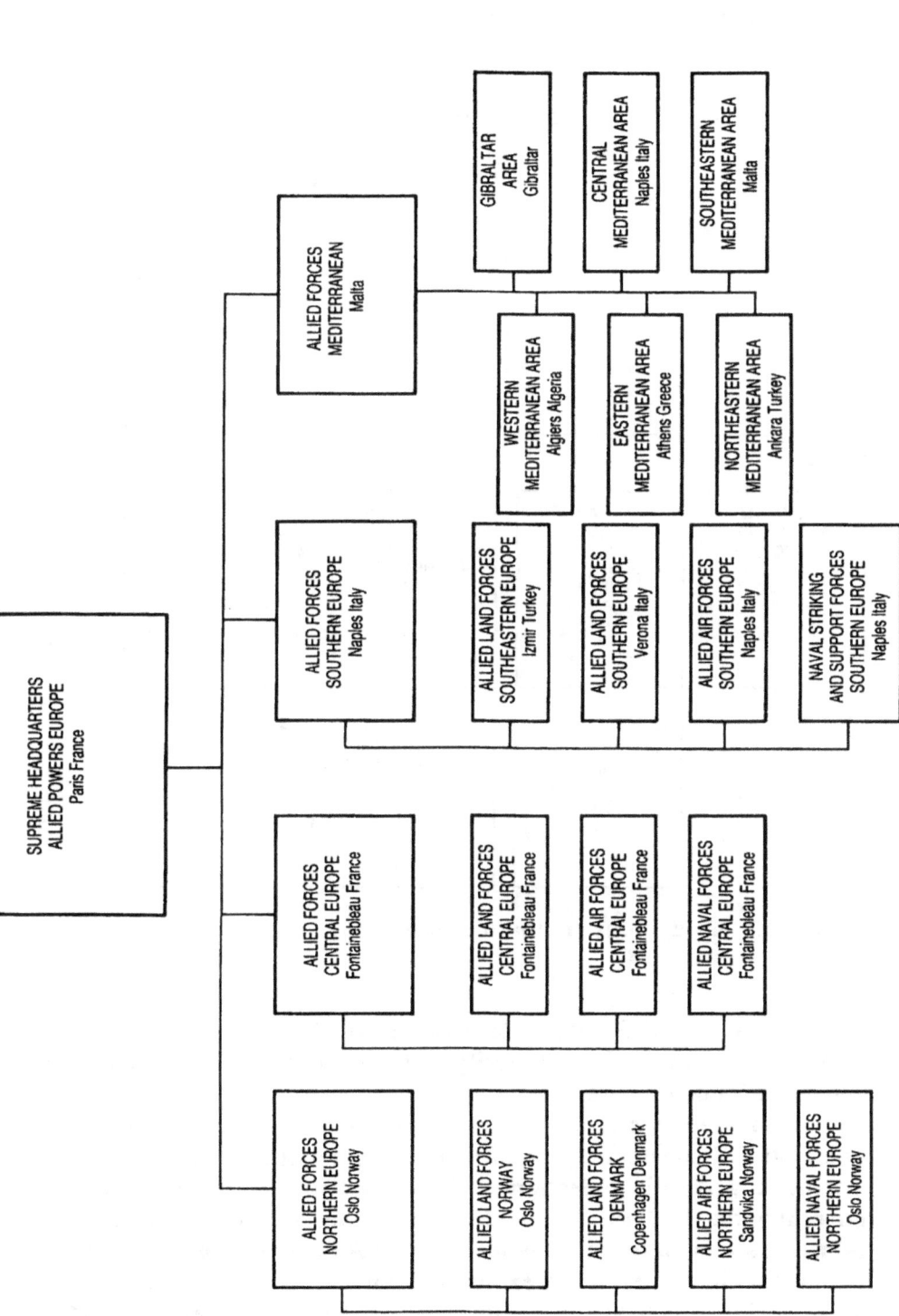

Figure 2.2 Allied Command Europe July 1954

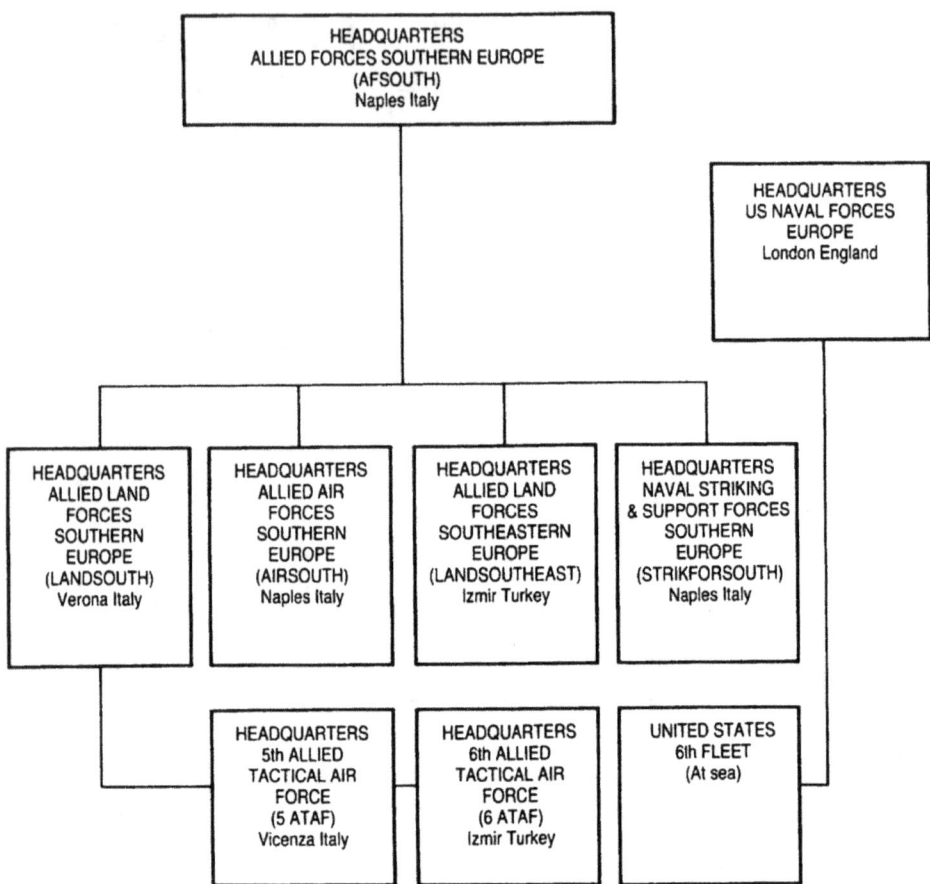

Figure 2.3 Allied forces southern Europe 1961

bombers of the US Strategic Air Command (SAC) over which the Alliance collectively had no authority, yet upon which the European allies depended for deterrence.

In later years, the issue of nuclear control within the Alliance would be a matter of considerable controversy. It is important to note, however, that there was little objection in the early fifties to the arrangements governing the Sixth Fleet in the Mediterranean. Not only were the European allies anxious to have this addition to NATO's nuclear underpinning, but the fact that it remained exclusively under US control was regarded as enhancing its deterrent value in Soviet eyes, precisely because the fleet could be used independent of a decision by the NATO Council. And indeed, during the Berlin Crisis of 1959, the US did place aircraft on fleet carriers on an advanced state of readiness and moved operations into the eastern Mediterranean. To be sure, the European allies remained uneasy about the exclusivity of US control of the Sixth Fleet, but they realized the deterrent value of this arrangement in peace even as they feared its consequences in a crisis.[45]

In addition to STRIKFORSOUTH, CINCSOUTH commanded LANDSOUTH (Italy), AIRSOUTH, commanding tactical air forces and LANDSOUTHEAST (Greece and Turkey). In the latter capacity he was responsible for control of the Turkish Straits.

CINCAFMED, with headquarters at Malta, was concerned with the use of the Mediterranean as a strategic avenue of communication in war. As with SACLANT, he had a multinational staff whose tasks were divided along functional lines: operations, plans, logistics, intelligence (commanded by a USN Captain) and communications. He had two deputy CinCs, one, a British Air Marshall, was responsible for air coverage of convoys moving through the Mediterranean. The other deputy CinC was an American Admiral, in addition to which there was a large American presence on the AFMED staff. Like SACLANT, CINCAFMED had no permanent forces. Contributing nations retained control of their earmarked ships and airplanes in peacetime. One problem which Mountbatten ran into was that coastal waters remained under national control, which in some cases extended 30 miles. He, however, recognized only 3 miles.

Subordinate area commands under CINCAFMED were given to a flag officer from that nation which had the longest coastline in a given area. Thus the French had COMEDOC (Commander Mediterane Occidental); the Italians, COMEDCENT (Commander Central Mediterranean); the Aegean and Eastern Ionian Sea under a Greek Admiral designated COMEDAST (Commander Eastern Mediterranean); while COMEDNOREAST (Commander Northeast Mediterranean), covering the Black Sea, the Sea of Marmora and the Eastern Aegean, was given to a Turkish Admiral. The southeastern Mediterranean from Malta to Cyprus including the Libyan and Egyptian coasts and the Levant were placed under the British CinC Mediterranean Fleet as COMSOUEAST (who was also CINCAFMED). The British naval commander at Gibraltar, who was already responsible for the region between Gibraltar and COMEDOC, was designated COMGIB.

Planning at AFMED was to include co-ordination with the national/regional commanders who sent permanent representatives to Malta. As noted, although the US did not officially come under CINCAFMED,

34 Seapower in the nuclear age

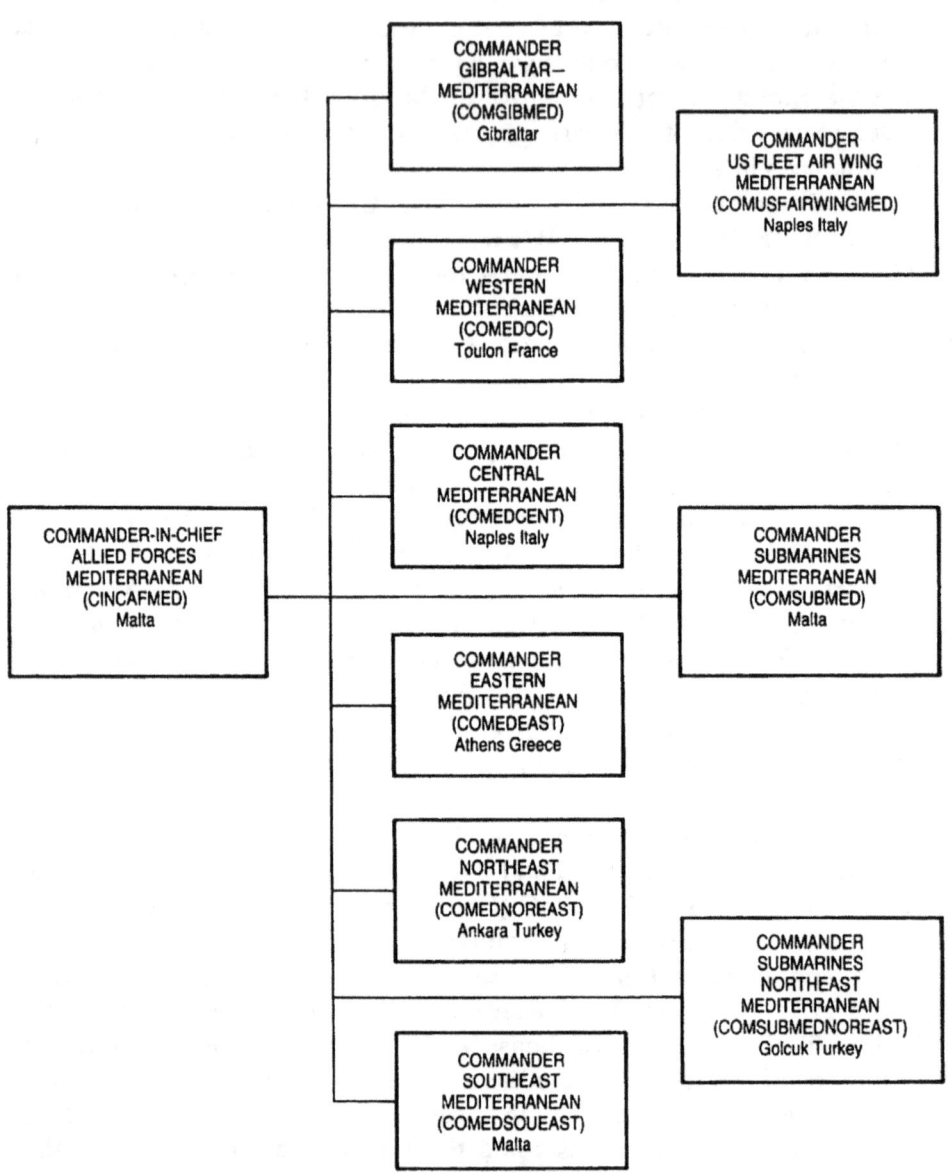

Figure 2.4 Allied forces Mediterranean 1961

American naval officers held staff positions under the Mediterranean Commander. Moreover, according to Mountbatten, in the event of war, US ships, other than those assigned to STRIKFORSOUTH, would enter the Mediterranean to assist in the war at sea. These would include submarines, hunter-killer groups, minesweepers and maritime air patrol squadrons (Mountbatten 1955: 176).

While CINCAFMED was responsible for air, as well as surface and subsurface protection of the SLOC, there were difficulties in obtaining commitments of naval air forces to AFMED. The French and Americans wanted to retain exclusive naval, as opposed to combined allied, control over their shore-based navy-owned aircraft. The Italians, Greeks and Turks maintained naval air forces as part of their regular air forces and thus they were subject to CINCSOUTH. The British land-based naval air belonged to the Royal Air Force, but in this case it was easier to co-ordinate with CINCAFMED, who was a British commander (ibid.: 179). It would not be until 1968, when the North Atlantic Council created COMMARAIRMED (Commander Maritime Air Forces Mediterranean), that the Alliance had a fully collaborative approach to naval aviation along the southern flank.

The first major large-scale naval exercises in the Mediterranean, MEDFLEXABLE and WELDFAST, took place in the fall of 1953 and 1954 respectively. Both were held in conjunction with CINCSOUTH. Other, smaller, AFMED exercises tested staff co-operation and, in general, the ability of the AFMED subordinate area commanders to implement AFMED war plans to secure the Mediterranean LOC (ibid.: 182).

AFMED planning was based upon an extremely favorable situation for the Alliance. In the 1950s and early 1960s there was no significant Soviet peacetime naval presence in the Mediterranean. To attack allied shipping, Soviet submarines would either have to come all the way round from Murmansk, through Gibraltar, or pass from the Black Sea through the Turkish Straits. The former route meant traversing oceans dominated by NATO navies, while in the latter case, passage in peacetime was governed by the Montreaux Convention requiring submarines to surface. Moreover, in war CINCSOUTH's ability to close the Straits was considerable.

The major threat to SLOC, as well as to the bases supporting allied naval forces, was regarded as coming from Soviet land-based air forces. A Soviet attack upon allied shipping was possible even if the Turkish Straits were held by NATO. According to Mountbatten:

> Enemy aircraft operating from Black Sea bases could carry out fighter-escorted strikes over the whole Aegean, light bomber strikes over most of the eastern Mediterranean basin, and heavy bomber attacks over the

whole Mediterranean sea. They could attack our ships with rockets, bombs and torpedoes, they could mine our ports and shallow harbours.

(ibid.)

The local air threat to convoys would have to be met by escort carriers. But AFMED's major protection from Soviet air attacks was not to come from any forces under his command. Rather it would be the American strategic air offensive, carried out by the strike forces of the Sixth Fleet, 'which will whittle down very considerably the scale of air attack which we would otherwise expect'.

In general, the primary strategic value of the Mediterranean during the first decade of the Alliance was the access it gave to NATO and Soviet territories for power projection. To be sure, securing the SLOC was a precondition for this use, but, without a major Soviet naval presence, the allies could almost be assured of immediate security of the seas along the southern flank. What would change in the late 1960s and early 1970s was not the strategic value of the Mediterranean to NATO, nor the measures taken to exploit that strategic value, but rather allied confidence in its ability to secure the seas quickly enough to support the power projection role. An indication of increasing NATO concern about security in the Mediterranean would be the reorganization of the naval commands, including the abolition of AFMED, the re-establishment of NAVSOUTH, directly under CINCSOUTH and the creation of a new command for naval aviation.

CINCHAN

The creation of the post of Commander-in-Chief Channel (CINCHAN) was part of the compromise involved in the appointment of an American as SACLANT. CINCHAN would be a British Admiral whose standing would be equal to that of SACEUR and SACLANT. In order to meet French reservations about placing the Channel under a British CinC, CINCHAN was to operate under the direction of the Channel Committee consisting of the naval Chiefs of Staff of Belgium, France, the Netherlands and the United Kingdom. The sea area under his command encompassed the English Channel and the southern North Sea, extending to the middle of the Jutland peninsula.

The naval forces earmarked for CINCHAN would be supplied by the nations sitting on the Channel Committee. As was the case with AFMED, the subordinate area commanders would be nationals from the coastal states of the sub-areas. CINCHAN's primary responsibility was to protect

Establishing the NATO Maritime Alliance 37

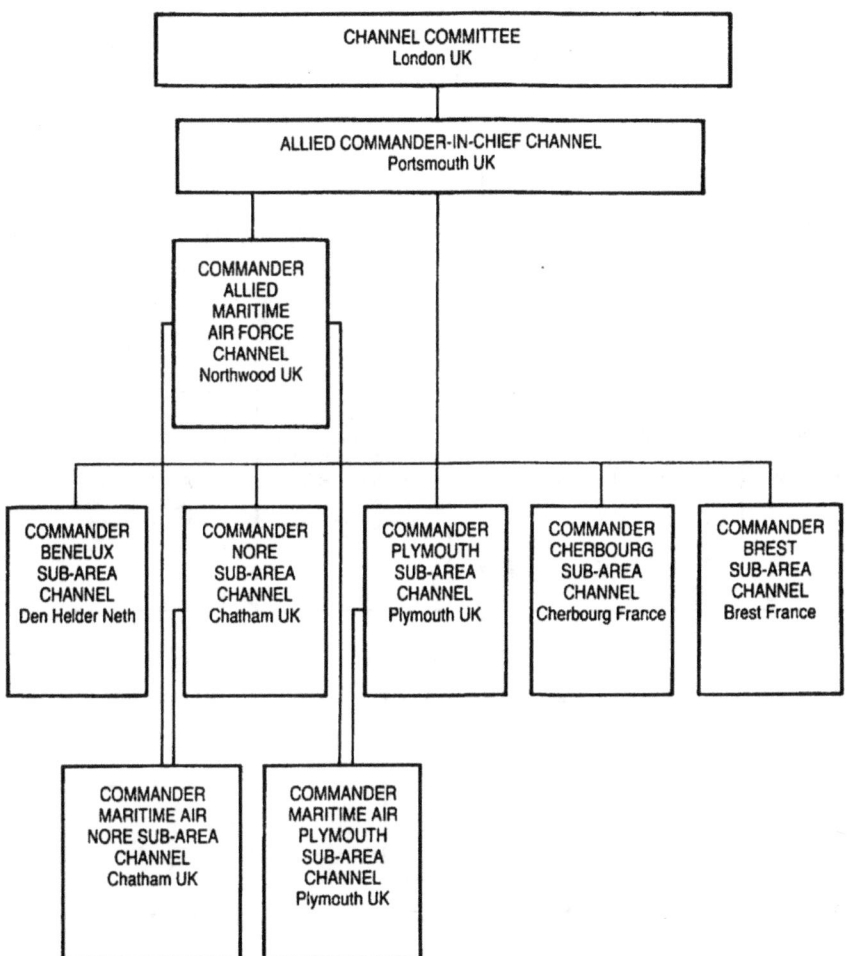

Figure 2.5 Allied Command Channel 1961

shipping in his area and secure the approaches to the European ports through which reinforcements would be moved.

SACEUR, CINCNORTH AND COMBALTAP

With the exception of the area covered by CINCHAN, SACLANT was to command all European coastal areas. In reality, SACEUR would exercise direct control over his immediate coastal approaches, excluding those covered by CINCHAN. Moreover, the smaller European navies

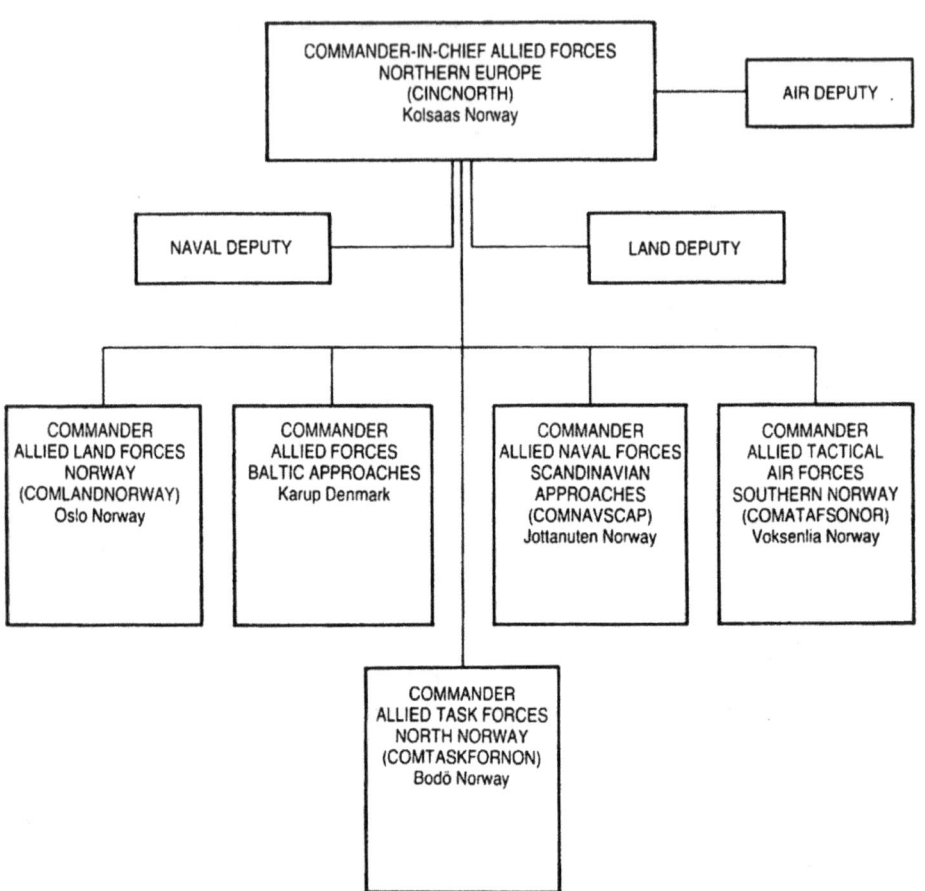

Figure 2.6 Commander-in-Chief allied forces northern Europe 1963

would earmark the majority of their forces to SACEUR's subordinate commands. While this did make for a somewhat awkward looking division of responsibility, especially in the North Sea, the naval command arrangements under SACEUR reflected more than a concern for national sovereignty and sensitivity.

One of the assumptions in the 1950s and early 1960s upon which NATO maritime strategy was based, was that attacks by Soviet submarines would be heaviest in the Baltic Sea and Scandinavian approaches. In the event of war, or during a crisis period, a premium would be placed upon early identification and destruction of Soviet submarines. Of even greater benefit to NATO would be the bottling up of the Soviet fleet at crucial choke points, especially the passage from the Baltic to the North Sea. At this time, the Soviet Baltic Fleet was the largest of its four fleets (the others being the Northern, Black Sea, and the Far Eastern). The Soviets were estimated to have 40–60 destroyers, 100–150 submarines and 800–900 naval aircraft, as well as minelayers and a dozen cruisers in the Baltic. These forces were said to pose a 'double threat', one, of a submarine offensive against allied sea lines of communication; two, of an amphibious assault in support of land operations against West Germany and Denmark (Jeschonnek 1956: 18). There were important Warsaw Pact naval facilities at Baltysk, Tallin, Kronstadt, Danzig and Konigsberg.

A Rand Corporation study reported at the time that the Swedes 'do not like the situation in the Baltic' (Robbins 1955: 60). According to senior Swedish naval officers, the Soviets had been engaged in 'feverish maritime activity' since 1950 giving them superiority in the Baltic. While the NATO powers would be able to contain Soviet submarines moving out on the high seas:

> In the inner seas and coastal waters, however, within reach of land-based enemy planes and near Soviet bases, the Western powers will have greater difficulties. Control of the seas here will always be fought over and in some cases partly surrendered to the Soviet.
>
> (ibid.: 58)

The 'minor naval powers' of the Atlantic Alliance would, according to these Swedish naval officers, have to keep their own territorial waters open through mine clearing, submarine hunting, protecting their shipping and in general keeping the sea lanes open to receive aid from the US. In view of the growth of Soviet naval power, Sweden, in the early 1950s, had initiated a major naval rebuilding program to include nearly 100 ships. Indeed, by the mid-fifties the only creditable local European challenge to the Soviets in the Baltic and Scandinavia could be mounted by neutral Sweden with its 18 destroyers, 27 submarines, 3 cruisers and 50 motor torpedo boats (MTB). Allied Denmark had only 6 submarines, 20 MTBs and 10 small frigates. The Navy of the Federal Republic of Germany was in the process of being rebuilt, as was the Norwegian fleet.

The rebuilding of the northern European allied navies was carried out in the context of general NATO planning, under SACEUR, to enhance security in the immediate coastal waters of individual states. Initially, both CINCNORTH, headquartered in Oslo, and CINCENT, in Paris, had naval subordinate commands, COMNAVNORTH and COMNAVCENT. COMNAVNORTH was responsible for the Norwegian coastal waters, thus including most of the Norwegian Navy, as well as the Baltic, which meant that German and Danish naval forces were earmarked to CINC-NORTH. COMNAVCENT covered that small area of the North Sea not covered by SACLANT, CINCHAN and CINCNORTH under the direction of a Dutch Admiral who had his own forces as well as German North Sea forces earmarked to him. The new German Navy was given primary responsibility for the defense of this portion of the southern North Sea. (The Germans voluntarily imposed a range limit on their naval forces of twenty-four hours sailing time from the approaches to the Baltic. It was not until 1980 that this limitation was removed [*New York Times*, 18 July 1980: A4].)

A 1956 study done at the US Naval War College by German Naval Captain (later Chief of the FRG Navy) Gert Jeschonnek outlined the tasks facing the German Navy in this region which were solely defensive. In the first instance, the Soviets had to be denied access to the Danish narrows in order to protect allied shipping moving through the North Sea to reinforce the northern flank. Second, since it was assumed the Soviets would try to supply their land armies from the sea:

> the German naval forces would have the further primary mission to deny the Soviet Baltic fleet the conduct of operations of this nature in order to prevent utilization of their sea lines of communication for supply purposes and to eliminate close support and protection to the Red Army's north flank by Soviet surface and sub-surface forces.
>
> (Jeschonnek 1956: 22)

In 1963 NATO reorganized its naval command structure directly under SACEUR with the creation of the Baltic Approaches Command (COMBALTAP). This new command was made immediately subordinate to CINCNORTH and had at its disposal in war the entire German Navy (on both sides of Jutland), the Danish Navy, the entire Danish Air Force, all of the FRG's naval aviation and some of its air forces, in addition to ground forces from the two countries.[46] This eliminated the need for an operational COMNAVCENT, a post which became merely one of naval adviser to COMCENT.

The officially stated purpose for the creation of COMBALTAP was to create a combined defense against Soviet submarines attacking allied shipping, and to prevent the movement of Soviet ground and supporting naval forces. NATO expected the Soviets to attempt to capture the 30-mile strip of land between East Germany and Hamburg as well as the Danish narrows, thus opening the way for safe passage of submarines into the North Sea. 'COMBALTAP,' according to a NATO communique, 'can expect air raids, paratroop drops and assault landings as well as sea bombardments.'[47]

The other components of CINCNORTH's naval responsibilities involved the immediate protection of the Norwegian coastal waters. Here there would be a great deal of overlapping with SACLANT's responsibilities which included protection of the sea lanes to the northern flank and projection of force ashore. However, since SACLANT's striking forces might well take time to arrive in Norwegian waters, it would appear that NATO wanted a command in the region which would initiate naval activities. The importance of the northern reaches of the North Sea would increase as the Soviets built up their northern fleet.

The naval command arrangements under SACEUR in the northern region suggest a number of considerations which were involved in allied maritime strategy. First, the Alliance did recognize that the Soviet Navy of the early to mid-fifties was essentially one with a limited range. It was anticipated that the major naval encounters would take place close to the European shore. Yet it was precisely in these immediate waters, particularly in the Baltic and its approaches, that the Soviet Navy could be most effective. Moreover, in the north, unlike in the Mediterranean, the Soviets would be able to bring their naval resources to bear in conjunction with their numerically superior conventional land forces. Second, and again unlike the situation which obtained in the Mediterranean, in the north NATO did not have overwhelming peacetime maritime superiority. The balance of forces directly adjacent to SACEUR's central and northern fronts favored the Soviets. American and British maritime forces would be able to secure the open seas for NATO, but SACEUR would have to rely initially upon the smaller European navies committed to him. Hence the importance placed early on in NATO upon establishing adequate naval command arrangements in the adjacent seas utilizing the available national naval forces.

The third consideration which appears to have informed the Alliance's maritime strategy in this region was the relationship between the land battle and the use made of seapower. As the combined nature of the COMBALTAP forces indicate, NATO was seeking to deny the Soviets use of the sea to support their own land campaign and at the same time secure the use of the seas for SACEUR. It was precisely because NATO anticipated a very fast-moving land battle that it took steps to insure adequate naval support. Such steps would become increasingly important as NATO moved closer to the official adoption of a flexible response strategy in the 1960s. As Gert Jeschonnek noted in 1969, the Baltic naval forces had to be available at all times, 'because there will be comparatively little margin for reactions in the form of a "flexible response" at the adjacent land front' (Jeschonnek 1969: 27).

Finally, the establishment of maritime commands subordinate to SACEUR and the earmarking of naval, as well as ground and air forces to secure the seas, again reflects concern about providing for the widest possible range of contingencies. In a war which quickly escalated to the strategic nuclear level, the coastal waters of northern Europe would be strategically important only as a base for the projection of nuclear force ashore. The maritime commands under SACEUR were primarily concerned with those tasks – protecting the sea lanes, denying the Soviets use of the seas – which would be significant in either a limited conflict, or in a more protracted struggle. Here again, since neither the political nor military leaders of the Alliance could predict how a crisis might develop, or what course an actual NATO–Warsaw Pact conflict might take, allied naval forces earmarked for the Baltic and northern coastal waters would have to put to sea prior to the outbreak of hostilities. Under the NATO command structure, both in the north and along the southern flank, with national forces essentially earmarked for duty off their own coasts, it would be likely that naval units would put to sea even prior to the formal activation of the allied naval commands and the assumption of operational control by the designated NATO commander. In many instances, such as COMNAVSCAP (Commander Allied Naval Forces Scandinavian Approaches), the allied commander was simply the national commander who was wearing two hats. The key point is that the arrangements established by NATO for the protection of the coastal waters directly adjacent to SACEUR were dependent upon no single conflict scenario. The strategic value of the seas would largely be determined by the nature of the land battle, its intensity and duration.

THE COMMAND STRUCTURE: AN ASSESSMENT

In many ways the maritime structure established by NATO to implement the Alliance's maritime strategy was a shell into which the moving parts would be placed in the event of war. Until 1968, not even SACLANT, the major naval command, had permanently assigned forces. In the Mediterranean, CINCAFMED was essentially dependent upon national naval commanders patrolling their own waters. He did not even have wartime operational control of the most potent fleet in his region of responsibility. Along the Baltic and Norwegian coasts, SACEUR depended for his initial naval support upon the German, Danish and Norwegian fleets being able adequately to meet any Soviet moves at sea in support of Soviet land

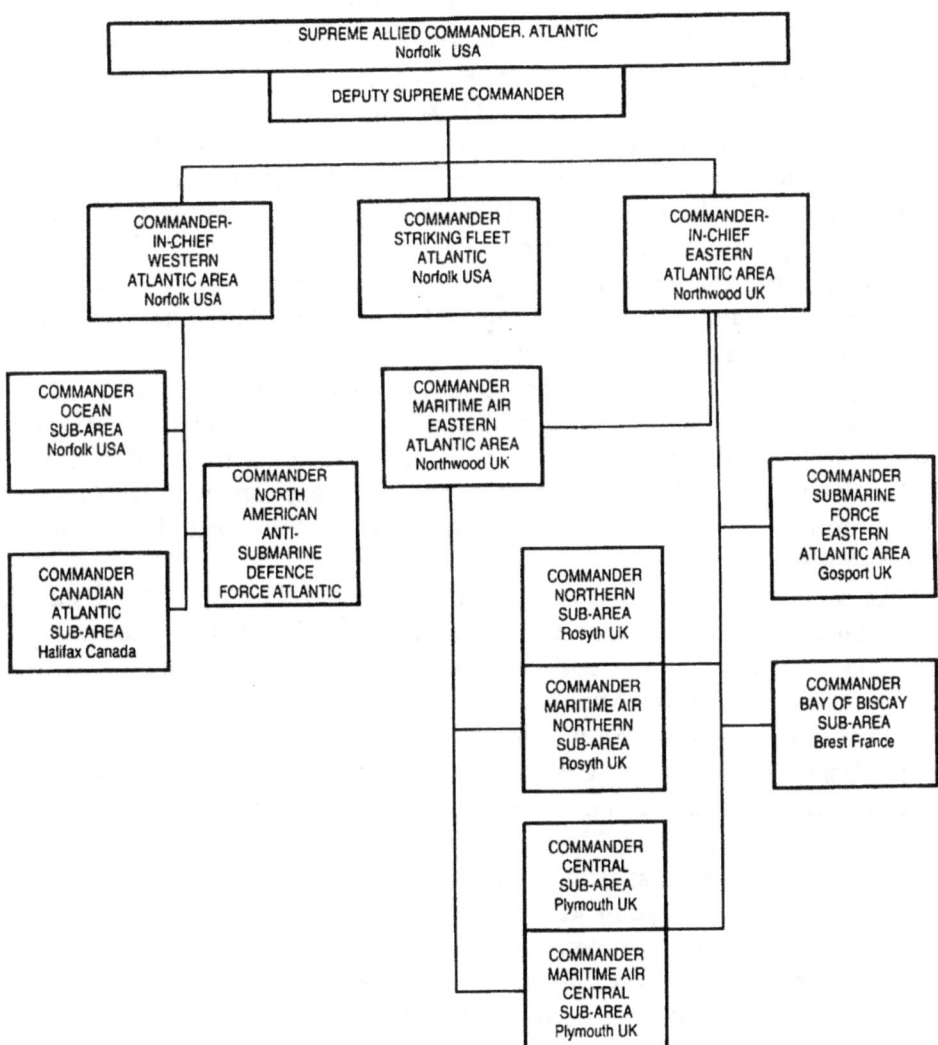

Figure 2.7 Allied Command Atlantic 1961

actions. Most importantly, the vast majority of allied seapower was American.

Not only was the maritime command structure a shell, it was a fragmented one. There was no single commander over all the allied maritime forces in war. SACEUR did have operational authority over the naval components of CINCNORTH and CINCSOUTH, but CINCHAN and SACLANT were supreme commanders on a par with SACEUR, and these two commanders were the national commanders of the largest fleets available to support SACEUR and owners of his Atlantic projection forces. Then there were the divisions of space. In the waters directly adjacent to the crucial central front, the sea was divided amongst the three NATO supreme commanders and several subordinate commands.

In comparison with the integrated forces on the ground in Europe, the organization of allied seapower was a mosaic of overlapping responsibilities and political compromises.

To point out these shortcomings in NATO's maritime organization is not, however, tantamount to saying that because of these shortcomings, the Alliance could not bring seapower into its overall deterrent posture, nor exploit the strategic value of the seas in a war, at whatever level, with the Soviet Union. The question is whether a lack of permanently assigned forces and the command divisions seriously affected the ability of the collective allied navies to perform their tasks in peace and war. Here, there are a number of considerations which mitigated to a large degree these shortcomings.

As mentioned above with regard to SACLANT, the NATO naval commanders were not Admirals without fleets. To be sure, they would assume command of the collective national forces only pursuant to a decision of the North Atlantic Council, but in the normal course of peacetime operations they would have considerable maritime forces at their disposal. This was because NATO naval commanders were also major national commanders. CINCHAN was simultaneously Commander of the British Home Fleet and Commander Eastern Atlantic under SACLANT. His air national commander was also Commander of Air Forces in the eastern Atlantic under SACLANT. Along the southern flank, the ASW Commander of the Sixth Fleet was also (after its establishment in 1968) Commander of Allied Maritime Air Forces in the Mediterranean.

Not only did this double- and triple-hatting mean that NATO commanders had actual forces under them, it also meant that there was basis for co-operation and co-ordination between national and NATO roles. For example, the ASW Commander of the Sixth Fleet would, in the normal course of his peacetime responsibilities, disseminate surveillance information to the navies of participating NATO nations in his NATO command. As one former Commander of the Sixth Fleet noted, this was actually a preferable situation since it provided for 'effective surveillance without duplication' (Kidd 1972: 28). In general, the fact that the NATO commanders would be performing important peacetime functions as national commanders implicitly gave the Alliance a permanent standing maritime contribution to deterrence. When NATO did establish and send to sea permanent combined forces, they consisted of no more than a half-dozen smaller ships, such as destroyers. While important for their symbolic value and in terms of providing some immediately available NATO naval force, they certainly were less credible than the simple fact that elements of the US and British fleets were continually at sea in the Atlantic and Mediterranean. And these were, for the most part, forces which were commanded by potential NATO commanders and earmarked for NATO use in war.

In the case of the smaller NATO navies, it simply made sense that their primary responsibilities under NATO commanders would be carried out in their coastal waters under national commanders who were also subordinate NATO regional commanders. Not only would this put to sea a force able to function in concert, it also meant that coastal surveillance could begin unilaterally without the operational wartime activation of the NATO command structure. This is aside from the fact that such an arrangement would be politically acceptable.

It should also be pointed out that while the major NATO commanders were also major national commanders, they did maintain separate staffs. Their NATO staffs were often multinational. Here again, however, there was a good deal of overlapping, with the NATO staffs being composed of the officers of their national command staffs for the same areas.

National commanders holding NATO commands was no firm guarantee that the national forces earmarked for NATO would automatically be made available in a crisis or war. National governments reserved the right to withhold their forces if they disagreed with the NATO decision to activate the combined naval commands. Thus, for example, Greece might not place its ASW destroyers at the disposal of AFMED, and restrict surveillance of Greek territorial waters by allied long-range maritime patrol aircraft. In this, however, the maritime situation was not entirely different from that which existed in other components of the collective forces of the Alliance. While it would be physically easier to withhold ships than, say, armored units stationed along the central front, NATO's maritime posture shared with the rest of its posture the conditionality associated with national decision-making and the voluntary nature of the Alliance itself.

There could also be some benefits in having national forces remain under national commanders until activation of NATO commands, especially with regard to sea-based striking forces, the carriers and ballistic missile submarines. These forces, mainly American, could be moved into position or put on alert status in advance of a collective decision by the

Alliance to activate its naval commands. In a fast moving crisis, this would be particularly important. Moreover, given that naval units were self-contained and could be shifted from one region to another more easily than ground forces, it would be possible for national contributions to be sent to where they were most needed. If the command structure was indeed a fragmented shell into which the national contributions would be placed, then there was no impediment to placing these contributions in one compartment or another, previous plans notwithstanding. In this sense the lack of permanently assigned forces would allow NATO to make use of the inherent flexibility of seapower.

From the standpoint of nuclear deterrence it would seem to make little difference that the maritime command structure was somewhat fragmented. In seeking to influence Soviet calculations as to possible NATO reactions and capabilities, the salient fact was that NATO could marshal, however organized, considerable sea-based nuclear projection forces in European waters. At the conventional level the same considerations would apply. For example, a Soviet planner would have to reckon with the fact that the British and Americans could move several carriers into the North Sea in response to a limited Soviet move against Norway, irrespective of the fact that the North Sea was divided between three NATO supreme commanders.

In terms of wartime co-ordination of forces, especially in the event of a protracted conflict involving a struggle for command of the sea, the fragmented command structure held the potential for serious problems. Ironically though, the national composition of the naval forces was expected partially to compensate for the area divisions. This was particularly the case between CINCHAN and SACLANT where

> it was hoped that the worst possibilities of operational misunderstanding at the command boundaries would be eliminated by ensuring that the same British officers who exercised certain British national commands and appropriate subordinate commands in Allied Command Atlantic (ACLANT), would also hold the subordinate posts in the Channel Command.
>
> (Hackett *et al.* 1978: 343)

Finally, it should be noted that until the late 1960s, NATO did not face a large number of permanently deployed Soviet naval forces. While it was anticipated that the main naval action would take place close to the European shore and that here the Soviets would have initial advantages, the Alliance did not face at sea the kind of overwhelming force it did on the ground in Europe. In order to be fully effective, even at limited ranges, the Soviet Navy would have to move out to sea, usually through choke points. Such movements would necessarily take more time than the mobilization and launching of ground and air assaults against NATO's land frontiers. If the Soviet Navy moved in advance of an attack elsewhere in preparation, this would be the signal for NATO to activate its naval commands and call upon earmarked forces. If the Soviets did not begin moving their ships until after the ground and air assaults had been launched, then NATO would still have time to put its maritime forces on war footing, relying on its ground and air forces to prevent a quick victory by the Soviets which would negate the utility of sea control and/or open the way for nuclear escalation. To be sure, there would not be much time available to the Alliance in any sudden surprise attack. However, since it would be likely that NATO would have some warning there would be time to activate maritime forces whose task it would be to secure the use of the seas. Since use of the seas for resupply and reinforcement would be important only in a protracted conflict, the lack of permanently assigned forces was not inconsistent with NATO maritime strategy.

With regard to the use of the seas for the projection of force ashore, both nuclear and conventional, the lack of a threatening Soviet force at sea meant that these forces could operate from a secure environment. That is, they could be moved into position and placed under a NATO commander without having to wait until the Alliance obtained local superiority. This again meant that NATO had some flexibility in marshaling its naval forces according to existing plans, or according to the prevailing situation in a crisis or war. In the case of nuclear strike forces, many of these were permanently on station under national command: for example, the Sixth Fleet and the three SSBMs assigned to SACEUR by the US, in addition to the British SSBMs. If an open conflict escalated quickly to the strategic nuclear level then neither the command structure nor the lack of permanently assigned forces could make much difference to the final outcome. All would depend on US strategic nuclear forces beyond NATO's control.

Thus, while not an optimal situation, the Alliance could hope to carry out its maritime strategy under the arrangements developed in the fifties because, with the exception of an immediate escalation to strategic nuclear war, there would be time to put ships to sea in order to secure the lines of communication and project power ashore. In later years, with a growing Soviet naval presence and advances in weaponry, this flexibility would be reduced.[48]

Chapter 3
The Cold War at sea: force and strategies

THE WESTERN EUROPEAN AND CANADIAN NAVIES

A 1959 report on the 'state of European security' put out by the Assembly of the Western European Union noted that naval forces would play a role in all contingencies, from nuclear war to limited war. In view of the growing nuclear parity between the Soviet Union and the United States, the report stated that emphasis should be placed upon conventional capabilities and command structures which would enable NATO to keep the sea lanes open by providing sufficient protection from Soviet submarines (AWEU 1959: 6–8). This continuing concern with security of the seas was reflected in the maritime force building programs of the Alliance members of the 1950s and 1960s. Although overshadowed by other aspects of rearmament, the naval construction of the smaller NATO members, and of the British and French, were to provide a large supplement to the American forces which dominated the Alliance's maritime posture.

The smaller European navies[1]

The primary task assigned to the Belgian Navy was to keep its vital ports open to allow the reinforcement and resupply of SACEUR. Thus, Belgium acquired 47 modern minesweepers. The Royal Netherlands Navy, whose primary task before the war had been the defense of Dutch territories in Southeast Asia, became almost totally dedicated to a NATO role. By the mid-1960s the Dutch had: 1 aircraft carrier (modernized in 1958), 2 cruisers (1 equipped to carry surface-to-surface missiles), 12 ASW destroyers (8 built since 1953), 6 ASW frigates (4 with helicopters), 12 submarines and 68 minesweepers.

Nearly all the Danish, Norwegian and German naval forces earmarked for CINCNORTH, particularly COMBALTAP, were newly built or refurbished by the mid-1960s. The Danes acquired several new destroyers and 6 submarines in addition to minesweepers and 4 new minelayers. Norway, whose immediate post-war Navy consisted of former British and Canadian ships, began a major rebuilding program in 1963 which included 15 submarines, 5 frigates, over 23 motor gun boats and 8 motor torpedo boats. The smaller craft were to be used 'to withstand the threat of the Russian trawler fleet which could be used for landing operations' (Kennet 1964b: 23).

As part of the German rearmament, the US and Britain supplied ships for the FRG Navy. By 1957 the Germans had 87 combat vessels and 111 auxiliary ships. Bonn had also begun a major naval construction program which eventually included 12 destroyers, 6 ASW fast frigates and over 15 submarines. Its Navy also acquired 79 Sea Hawk fighters from the US.

Along the southern flank, the Italian Navy had over 40 destroyers by the early 1960s, as well as 3 cruisers including the missile-firing *Garibaldi*. The other two smaller navies in the region, the Turkish and Royal Hellenic Navies, continued to rely on former US and British ships. Turkey had 9 destroyers, 10 submarines, 13 minesweepers, 10 coastal escorts and 30 smaller patrol vessels, while the Greeks had only 1 cruiser, 6 destroyers, 2 submarines, 5 minesweepers, 2 minelayers and various smaller ships. The relatively weak economic condition of these nations appears to have precluded a major naval rebuilding program by either. In the south at this time, the British maintained a fleet earmarked to AFMED for protection of the Mediterranean SLOC which, combined with the presence of the Sixth Fleet on permanent patrol, made up for the small contributions from the two Eastern European allies.

In the 1950s the Portuguese Navy was expanded and modernized under the US Military Defense Assistance Program, so that by 1963 it included 14 escorts, 3 submarines, 18 minesweepers as well as about 50 smaller coastal units (IISS 1963: 20). During the 1960s, the French and the Dutch jointly financed the construction of new vessels. By the end of the decade Portugal had 13 new frigates, 4 ocean minesweepers, 4 submarines and 13 coastal escorts (IISS 1968: 28). Throughout this period, Portugal's major military activities, including those involving her Navy, were directed towards counter-insurgency in her African colonies. Only three vessels were earmarked for NATO although the Portuguese Air Force did earmark a squadron of American P2V5 Neptunes for anti-submarine and reconnaissance tasks in the IBERLANT region. (Following the collapse of Portugal's African empire in the late 1970s, the Navy was brought back home and assigned NATO-related tasks under IBERLANT.) The most important contribution Portugal made to NATO's naval posture was to make available to the US and France, on a bilateral basis, and to NATO as part of the infrastructure program, air and naval bases on the continent and the Atlantic islands, as well as providing locations for fuel depots, ammunition storage and communication facilities (Crollen 1973: 61–5).

France

The French Navy, the second largest of the European navies, was originally dedicated to sea lane and coastal protection in European waters in the Atlantic and the Mediterranean. With the coming of De Gaulle and the Fifth Republic, France not only sought to reshape its naval forces to provide for more projection capabilities, including nuclear projection, it also initiated a staged disengagement from the NATO naval command structure, one that preceded its later withdrawal from the Alliance's integrated military system.

As part of its effort to create a more balanced fleet, France built two new aircraft carriers, the *Clemenceau* and the *Foch* and equipped them with its own Dissault Etandard IV and the US Crusader aircraft, the former capable of delivering the atomic bomb. Work also began on an SSBM to carry a French-built missile. In addition, France re-equipped its destroyers, many built since the war, with advanced anti-air and anti-submarine weapons. Several guided-missile frigates, carrying ASW helicopters and general purpose ASW frigates, were built between 1957 and 1962. A program of submarine construction, also begun in 1957, produced nineteen modern conventional ocean-going vessels.

The development of the French Navy into one of the most modern in the world coincided with the progressive disengagement of France from NATO naval commands. In 1959, De Gaulle withdrew the French Mediterranean fleet from AFMED's earmarked forces. Part of the explanation for this was that with important North African interests, France, in the event of war, would be concerned with maintaining the north–south SLOC, whereas AFMED's interest was in west–east lines of communication (Mulley 1962: 169). There was also French concern with the on-going war in Algeria. In general, this decision regarding contributions to AFMED was a tangible affirmation of France's desire to achieve a more independent military policy. It did not, however, mark a complete disengagement from allied efforts in the Mediterranean. NATO did maintain use of the French naval facility at Mers-el-Kebir, Algeria (location of a storage depot under the Alliance's infrastructure program, a summary of which follows below). Moreover, French officers continued to serve on AFMED's staff and a French Admiral remained COMEDOC.

In 1960, France asked that a proposed new SACLANT subordinate command, IBERLANT, covering the western approaches to the straits of Gibraltar from the Bay of Biscay in the north, west to the Azores, and south to the Dakar region, be placed under a French Admiral. The NATO Council would not agree to this unless the previous French decision to withdraw its Mediterranean fleet was reversed. The French did not comply.

Three years later, in June 1963, De Gaulle announced that from French naval forces in the Atlantic earmarked for SACLANT, some sixteen destroyers (as well as an anticipated reinforcement of cruisers) would be withdrawn (*The Times*, 21 June 1963: 12). This was followed in April 1964 by the withdrawal of all French officers from the staffs of CINCHAN and AFMED. The major NATO naval exercise that fall, OPERATION TEAMWORK, the largest exercise to date, did not include French forces for the first time since the establishment of SACLANT (de Cormoy 1970: 305). In 1966, the NATO Council established the IBERLANT command with an American Admiral, headquartered at Lisbon in command of earmarked British, Portuguese and US forces. That same year, as part of the overall French withdrawal from NATO's integrated command structure, a French Admiral ceased to be COMEDOC under AFMED. (AFMED itself, however, was abolished a year later.)

It is difficult to assess the exact impact of the French disengagement from the NATO naval structure. For planning purposes, NATO could no longer count upon additional ships in the Atlantic, the English Channel or the Mediterranean. With a growing Soviet naval presence off the southern flank, there was concern that the French withdrawal would have the effect of isolating Italy and the other southern flank nations from the rest of the Alliance (see AWEU 1967: 8). There was also the problem of NATO now having no, or limited, access to French ports and infrastructure facilities.

But it remained an open question as to how France in general, and her maritime forces in particular, would respond in the event of a war or crisis. After all, the naval forces, like those of all other Alliance members, had only been designated for NATO use. Nothing prevented France from contributing them in the future or of continuing joint planning. At the time of the 1963 withdrawal from the Mediterranean and the English Channel a senior French Admiral seemed to go out of his way to assure the allies that his government's decision would not undermine collective defense. Pointing to the situation in the Mediterranean, he noted that:

> in practice nothing has changed since 1959 as regards the relationship of the French navy with the allied navies and co-operation with them

is still close and constant in planning and joint exercises. There can certainly be no question of any allied navy planning its own 'private war'.

In this 'narrow sea' war can only be total and 'the interests of France are inseparable' from those of NATO.[2]

In the summer of 1964 the French Commander-in-Chief Atlantic (CECLANT) and SACLANT concluded an agreement of co-operation. In his 1967 annual report, USCINCLANTFLT reported that the agreement 'is currently accepted as satisfactory assurance that France will conduct NATO missions in the western English Channel, Bay of Biscay and IBERLANT' region. However, he did note that 'on occasion when France does not co-operate, such as withdrawal of French forces from Exercise Teamwork, the question of "good faith" regarding SACLANT/CECLANT arises'.[3]

While such reservations did exist amongst allied naval commanders, the pattern of co-operation in the Atlantic was repeated in the Mediterranean on a bilateral and trilateral basis as well as on a French/NATO basis. Shortly after the 1959 withdrawal from AFMED, an American, British and French agreement designed to 'co-ordinate search and tracking operations on the transits of specially designated Soviet/satellite units ... was activated during the transit of four Soviet Whiskey class submarines from the Baltic to Vlone Bay, Albania'. USCINCNELM reported 'excellent results' from the combined operation.[4]

In the years following the 1966 French withdrawal from all NATO integrated command structures, co-operation between French and allied naval forces continued. A French Admiral was posted as liaison to SACLANT at Norfolk. According to a 1973 memo from COMNAVSOUTH to CINCSOUTH, cryptographic key material for ship broadcast, employed by NATO's naval communication agency, was 'routinely' being released to the French.[5] A 1976 report to the Assembly of the Western European Union, *Security in the Mediterranean*, noted that French co-operation with COMNAVSOUTH was 'reasonably close'. French naval vessels regularly participated in joint exercises and maritime aircraft were also co-operating with COMMAIRMED. In addition, NATO and France had developed various joint 'contingency plans' in the event of a crisis or war in the Mediterranean (AWEU 1976).

It would appear, therefore, that the withdrawal of earmarked French naval forces from the NATO commands was to a large degree compensated for by the continuation of co-operation by other means: bilateral arrangements with the US, frequent joint exercises, the maintenance of ties with the Alliance through informal and formal agreements. Any predictions as to the reaction of French naval forces in the event of war or crisis would, on these grounds, weigh heavily in favor of a positive French response to the mobilization of Western seapower. Indeed it is almost impossible to imagine that if the seas really did become crucial to the defense of Western Europe in a conflict with the Soviet Union, France would hold back its naval forces. At the very least, she would have to patrol her own waters and this would be of benefit to NATO. As to the French sea-based nuclear forces, there would seem to be no scenario other than strategic war in which these would be employed. As such, the question of earmarking for NATO is of little consequence since they would not make much difference, and in any case would remain under national command. It could well be that the French have co-ordinated their sea-based nuclear forces with the US and the British, again mitigating the adverse implications of France's *de jure* withdrawal from NATO's maritime posture.

United Kingdom

The establishment of the NATO command structure and the development of seapower directed towards the defense of the north Atlantic region coincided with the retreat of British seapower from its global position. For the Alliance, the withdrawal of the Royal Navy from its far-flung responsibilities was fortuitous since it eventually meant that 'virtually the whole of the Royal Navy – the largest and best equipped NATO navy in Western Europe – was earmarked for assignment to the Alliance'.[6]

In addition to the decline in British interests overseas, the atomic bomb and the expectation that any future central war would be short raised questions about the continuing utility of seapower. Similar doubts existed with regard to the US Navy but America, unlike Britain, was a global power on the rise and it did have the atom bomb. Defense White Papers in the 1950s sought to define a role for the Navy in the new age. It was the British who introduced the concept of a 'broken-back war', wherein navies would be important after the initial nuclear exchange, to carry on the conflict. The White Papers also made reference to the more traditional uses of seapower: searching out and destroying enemy ships, protecting lines of communication necessary to support 'our warlike operations and to safeguard the supply lines of the allied countries' and 'to provide direct air support for operations ashore in those areas where it could not be given by land-based operations' (Hampshire 1975: 116).

By the early 1960s, despite some uncertainty regarding the proper role of the Royal Navy, the RN had a fleet of some 284 ships. Although this represented a reduction in absolute numbers, the fleet was being significantly upgraded for conflict at sea short of strategic nuclear war. This included new destroyers, some carrying guided missiles, the deployment of a new generation of carrier-based aircraft on the four carriers in service and, between 1953 and 1963, the construction of 36 frigates (3 anti-air, 3 air direction, 11 general purpose, 19 ASW). During this period the RN also acquired 8 new conventional submarines (AWEU 1963: 9).

Under the terms of the Nassau agreement between the US and the UK, Britain acquired the American Polaris missile, thus giving the Royal Navy a role in England's strategic nuclear posture. According to a 1963 statement on defense, the RN had been 'entrusted with a most important additional task', that of operating 'a force of Polaris equipped nuclear submarines as Britain's independent contribution to the long range strategic forces of the Western Alliance' (Amme 1967: 162). It would appear that the British force, unlike the French SSBM and carrier-based nuclear force, was earmarked for NATO, although like the American contribution, remained under national command and would do so in the event of war. According to a 1978 UK report, the Polaris force 'which maintains a continuous patrol at sea, is the only European contribution to the NATO strategic deterrent' (UK Ministry of Defence 1978: 12).

Canada

A 1968 official history of postwar Royal Canadian Navy (RCN)[7] observes that: 'After signing the North Atlantic Treaty in April 1949, Canada decided that her naval contribution to the allied military organization should take the traditional form of anti-submarine forces, which would be available to specific NATO commands for definite tasks' (Canada, Dept of Defence 1968: 4). Shortly after his appointment as the first SACLANT, Lynne McCormick was told by the Canadian government that Canada would concentrate its contribution to the Atlantic Command in the area of convoy escort.[8]

The Canadian naval rearmament of the 1950s reflected the early allied concern with ASW and especially the need for convoy protection in the event of war. By 1956, Canada had built or converted fourteen ASW destroyers and had dedicated its lone carrier to the convoy escort role. It had acquired long-range as well as carrier-based maritime patrol aircraft. By the time the RCN's naval expansion had been completed, it had a total of 154 ships of all kinds, with a majority of the major combatants located in the Atlantic and earmarked for SACLANT (Dillon 1972: 17–20).

During the 1960s, the number of Canadian ships declined sharply, but, as with the British and French navies, significant improvements were made in ship capabilities. Destroyers were refitted with advanced ASW equipment and a new class, the Tribal Class, was launched carrying the Sea King ASW helicopter. Canada acquired the US ASROC anti-submarine rocket-boosted torpedo. Prior to the major defense cutbacks initiated by Prime Minister Trudeau in 1968, the Canadian contribution to allied defense in the Atlantic included: 1 light ASW escort carrier, 9 ASW helicopter destroyers, 3 destroyer escorts, 2 submarines, 3 squadrons of Argus long-range ASW patrol aircraft and 24 Tracker ASW aircraft (flown from the carrier) (Canada, Dept of Defence 1968 and 1964).

In peace and in the event of war, Canada would be responsible for its own coastal waters and an additional segment of the North Atlantic, including the Davis Strait. Both Canada and the United States were part of NATO's North American region which, although within the territory encompassed by the North Atlantic Treaty, remained exclusively a bilateral concern. (There was a Canada/US Regional Planning Group.) Over the years, through agreement with the US Navy, the NATO chain of command became identical to the chain of command which would go into effect in the event of a bilaterally declared emergency. Thus, if the US and Canada both agreed that the situation merited an alert of their Atlantic forces, Canada's ASW forces would be placed under Commander North American Anti-Submarine Force, an American Admiral, who was also the Commander of SACLANT's ASW Force (Canada Dept of Defence 1964: 3).

Continual surveillance of the seaward approaches to North America, and heightened surveillance during crisis periods, became increasingly important as the range of Soviet submarines increased and as the Soviets developed SLBMs capable of firing upon targets in the US. In this sense, Canadian co-operation with the US, and the meshing of this arrangement with NATO responsibilities, was another indication of how allied naval tasks complemented national and regional responsibilities. For example, during the Cuban Missile Crisis, North American forces were placed on alert. As CINCLANTFELT reported, the American Caribbean build-up required additional surveillance. 'Canadian Argus aircraft under CAN-COMAIRLANT increased their ASW surveillance and their assistance and co-operation throughout the crisis contributed significantly to the ASW effort.' 'Without this valuable assistance,' he noted, 'much of the Western Atlantic area would not have been adequately covered.'[9]

NATO INFRASTRUCTURE SUPPORT AND COMMUNICATIONS

From the preceding it is evident that while the NATO maritime commands did not have permanently assigned forces, the individual navies which earmarked forces for the Alliance, and the French Navy which sustained its co-operation with NATO maritime commanders, did seek to maintain adequate forces to be able to perform national and NATO roles. One major problem, however, was that with the NATO maritime commanders relying upon earmarked forces, they also had to rely upon national logistical support. Each nation was responsible for the resupply and refueling of its ships at sea and earmarked naval air forces. Although SACLANT did test some multilateral at-sea resupply procedures, he and the other naval commanders remained dependent upon national resupply of individual ships.

Here again there were a number of factors which mitigated, to a certain extent, the potential weakness of national logistical support. First, the smaller European navies under CINCNORTH were not tasked for 'high-seas' duty, and therefore their range of operations would keep them close to their home ports. This would also apply to the Greek and Turkish navies in the south. Second, the larger European navies, the Italian and especially the Royal Navy, did maintain at-sea resupply capabilities. The French Navy, developed as an independent force, included resupply and refueling vessels. Third, by far the largest of the NATO navies, and the one upon which the Alliance's entire maritime strategy depended, the US Navy, maintained an extensive support structure for its forces forward based in European waters. The United States also maintained a global communications and surveillance system which could be placed at the disposal of NATO. For example, in the early 1950s the US had begun to deploy lines of hydrophones on the ocean floor off its Atlantic and Pacific coasts, as part of the Sound Surveillance System (SOSUS). Later, similar lines were deployed on the floor in the vicinity of the Greenland–Iceland–United Kingdom gap and off the Norwegian coast in order to help identify submarine movements and, if necessary, direct killer forces to particular sea areas (Friedman 1980: 120; SIPRI 1970: 148–54).

Finally, the European navies, but especially the USN, were partially supported by the Alliance's infrastructure program. For maritime strategy this meant the construction or improvement of naval airfields, port facilities and storage depots in member countries drawing upon NATO general funds. This was a slow process, and American commanders often complained in their annual reports of the lack of progress in the NATO naval infrastructure program.

By the late 1950s, a number of special (atomic) ammunition storage facilities had been built including one at Hvalfjorihur, Iceland, while a large NATO underground warehouse to be used by the French, Italian and US navies had been opened at the Algerian French naval facility, Mersel-Kebir. (The 1959 French withdrawal from AFMED did not affect the base's status. Its use was lost when Algeria gained independence. With the French withdrawal in 1966 the USN and NATO lost use of naval facilities at Brest and other POL [petroleum, oil, lubricant] facilities and airfields in France.) Work had also begun on underground dry food and torpedo storage depots in Turkey.[10] In 1959, the US Navy, for the first time, took permanent occupancy of a NATO maritime airfield in the Mediterranean, at Signonella, Sicily[11] (although CINCNELM reported that the USN had to make a number of improvements to bring it up to operational readiness). Work was also begun on improving the air facilities at Souhda Bay, Crete, Lisbon, Brest and on the Azores. A CINCLANTFLT report in the early 1960s notes that 'POL and logistic support posture for general war was further improved by the activation of NATO storage depots in Scotland and England'.[12] By 1967, the Alliance had some sixty-five naval-related infrastructure projects still in the process of being completed.[13]

In the mid-1960s, SACLANT also tested, on a limited basis, the ability of the Alliance to support combined forces at sea for an extended length of time. The MATCHMAKER exercises involved roughly a half-dozen ships from several nations on continuous patrol for up to six months, co-operatively supported through the Alliance. By 1968, NATO had created its permanent standing force, STANAVFORLANT (Standing Naval Force Atlantic), which did avail itself of infrastructure support, but whose individual ships also relied heavily upon national logistical support.

Other co-operative measures taken in the maritime area included the establishment in 1951 of the Allied Naval Communications Agency (ANCO). Responsible to the Military Committee, ANCO was to meet the requirements of the major NATO commanders for adequate and reliable communications for maritime operations. In order to improve Alliance-wide ASW technology, the SACLANT Anti-submarine Warfare Research Center was established in 1962 in La Specia, Italy.

Overall, it would appear that logistics and the general ability of the various national maritime forces to work together remained a serious concern for allied naval leaders, particularly the American commanders. Not only was progress slow in the construction of new facilities, but the withdrawal of France from the integrated command structure sharply curtailed NATO peacetime access to facilities in that country. Yet, while the existing support and communications structure fell short of what would have been needed to wage a protracted war, it went well beyond anything which had existed between allied nations in peacetime. Together with the development of national maritime forces, allied co-operation in various undertakings did make NATO a maritime alliance.

NUCLEAR WEAPONS AND THE MLF

While most of the allied force developments were in the area of non-nuclear or general purpose maritime forces, nuclear weapons were an important element in the Cold War allied maritime posture. They were deployed for projection purposes and for use by allied forces at sea to secure and/or deny use of the sea for the Alliance.

The US Navy had been quick to realize the potential of nuclear weapons in enhancing the effectiveness of (and securing continuing funding for) its aircraft carriers. By early 1951, non-nuclear atomic bomb components were deployed on USN carriers in the Mediterranean. The nuclear components, which remained under the civilian control of the Atomic Energy Commission, were to be flown to the carriers in the event of war. Assembled on the carriers, the nuclear bombs were to be delivered by the new AJ-1 and P2V-3C aircraft. By the end of 1951 'the entire set of nuclear and non-nuclear bomb components were deployed on several American carriers'. Targets included Soviet submarine bases and other military facilities within a 600-mile radius of the Mediterranean (Bracken 1983: 78–9).

This deployment coincided with the beginnings of the build-up of land-based tactical nuclear weapons by the American forces in Europe. It was also in line with the increasing emphasis on nuclear weapons of the Eisenhower administration. In July 1954, Supreme Headquarters Allied Powers Europe (SHAPE) formally recommended that tactical nuclear weapons be integrated into the allied defensive posture. In December of that year the North Atlantic Council agreed to the plan which brought NATO strategy into line with the 'New Look' American strategy. The Supreme Commander was authorized to assume that nuclear weapons would be used in a conflict with the Soviet Union. Field Marshall Bernard Montgomery stated that SHAPE was basing all its operational planning 'on using atomic weapons in our own defense ... whether the aggressor has used them or not' (Enthoven and Smith 1971: 120).

In the 1950s, the allied use of atomic weapons was viewed as necessary to support conventional ground, air and sea operations in the defense of Western Europe. The US Strategic Air Command's (SAC) Emergency War Plan (EWP) listed the objectives of the atomic offensive as: the blunting of the Soviet capability to deliver an atomic offensive against the United States and its allies, the disruption of the vital elements of the Soviet war-making capacity and the retardation of the Soviet advances in Western Europe as close to D-day frontiers as possible. Included in this latter objective was securing allied lines of communication (Rosenberg 1981/2: 9–12). Lines of communication for reinforcement and resupply continued to be important because the atomic offensive against advancing Soviet forces required that allied ground and air forces be able to blunt a Soviet attack, compelling the Soviets to concentrate their forces, thus making them easier targets for atomic weapons (ibid.: 12).

In the earlier EWPs, 'naval' targets, those selected by the US Navy, were not formally listed. Carrier task forces retained the freedom of action to attack targets as the tactical situation required (ibid.: 8, FN13). It is evident that the USN believed it could best secure the sea lines of communication through the execution of atomic strikes against Soviet port facilities and naval bases. The USN would carry out such strikes in conjunction with conventional maritime operations undertaken by its own and allied general purpose forces. A US Defense Department Weapons Evaluation Group (WSEG) report of February 1955 on the 'Evaluation of an Atomic Offensive in Support of the Joint Strategic Capabilities Plan' noted that:

> With respect to the objective of securing Allied lines of communication, it appears that if planned Allied efforts to counter Soviet threats to sea transport are implemented, Allied merchant shipping losses will not cause a critical reduction in the support of the overseas theaters of operations. In addition to limiting Soviet submarine operations, the US atomic strikes against Soviet naval bases and supporting facilities are expected to reduce substantially the Soviets' capability to conduct amphibious operations and to provide seaborne logistic support for their forces.
>
> (ibid.: 36)

As noted above, General Eisenhower had already received assurances that US aircraft carriers assigned to the Sixth Fleet would be available to SACEUR. By 1954, American aircraft carriers, earmarked for SACEUR, had begun to shift their primary focus from battlefield support to nuclear strikes against targets deep in Warsaw Pact territory, including the Soviet Union. Upon receiving an alert signal from SACEUR, the fleet would come under his command as STRIKFORSOUTH (Allied Striking Force South). The USN would then prepare to execute its long-range nuclear attack plans (Dur 1976: 66).

Throughout the 1950s, the US military services, particularly the Navy and Air Force, argued over target responsibilities. Despite various efforts at co-ordination, duplications and triplications (two or more commands delivering nuclear weapons to the same target) were not reduced, especially in the European theater (US Dept of Defense n.d.). In 1959, representatives of SAC, USCINCEUR (Commander-in-Chief US Forces, Europe) and CINCLANTFLT withdrew from targets already covered by SAC. In reporting this withdrawal, however, CINCLANTFLT noted that his plans continued to dictate the employment of nuclear weapons by the Sixth Fleet and other American forces earmarked for NATO.[14] A year later, in April 1960, NATO held its first target co-ordination conference involving SAC, SACLANT and SACEUR.[15] With the development in 1960 and 1961 of the Single Integrated Operational Plan (SIOP), US nuclear strike forces, including those of the Navy, were formally co-ordinated and conflicts sharply reduced.

By the time the SIOP was adopted, the USN had significantly increased its capacity to support SACEUR with nuclear weapons. In his 1957 report, US CINCLANTFLT noted that: 'Atomic delivery capability due to increased allocations and the authority to disperse 100 per cent of the allocation has doubled in the past year.' As SACLANT, he had instituted an atomic weapons course for allied officers and US personnel attached to the Sixth and Second Fleets and had begun target training for those targets assigned to these naval forces. A year later, he reported that the dispersal of atomic weapons had increased by 30 per cent, and components by 50 per cent. New storage facilities had been built into the carriers *Franklin Roosevelt* and *Saratoga*. Combined with the deployment of A3D and A4D aircraft, this meant that the 'capability for delivery of atomic weapons has increased dramatically'. A procedure known as the 'Daisy Chain' was developed to provide for the movement of atomic weapons from the US to the Sixth Fleet.[16]

A review of NATO exercises at this time indicates that SACEUR relied on the carriers for support in his southern region. For example, during exercise CRESCENT MACE in 1959, Sixth Fleet aircraft began atomic strikes against 'all SACEUR and CINCSOUTH target programs' within eight hours of the 'R-Hour message'. 'Atom play' was also used for amphibious landings in northern Italy and to support COMLAND-SOUTHEAST forces.[17]

In his 1961 report the US Commander-in-Chief Naval Forces Europe (CINCUSNAVEUR) noted that:

> In co-operation with CINCSOUTH, COMSTRIKEFORSOUTH representatives at the Joint Command and Operations Centers (JCOC) at Verona Italy and Ismir Turkey have simulated full play of CINCSOUTH regional program utilizing NATO strike communications nets.
>
> As the fiscal year drew to a close, an optimum 2 CVA SACEUR scheduled Program Atomic Strike Plan, which is co-ordinated with SIOP, was effective. It covers all significant threat targets in the area of interest to the Sixth Fleet. The plan has been developed on the basis of immediately responsive alert posture without unduly restricting normal fleet operations throughout the Mediterranean.[18]

At the time of the 1959 Berlin Crisis, the fleet put both its sea control and nuclear strike forces on alert. From 4 May to 2 June one carrier task unit was maintained at sea at all times and for most of the period both task forces assigned to USCINCEUR and SACEUR were at sea. Each carrier maintained at least six aircraft in 'immediate launch status'.[19] Two years later, amid the 1961 Berlin Crisis, the Sixth Fleet was again put on alert and reinforced with additional long-range ASW aircraft.[20]

Though these measures were taken unilaterally by the United States, they were essentially the same as would have been taken had the Alliance moved collectively to place its naval forces on alert and under allied command. A study prepared by the Center for Naval Analyses noted with regard to the 1959 alert that: 'given that the 6th fleet CVA (attack carriers) force still had a major role in SACEUR's scheduled program, this was an important political signal to the Soviet Union, an indication of US determination'.[21]

In May 1962, NATO's sea-based nuclear strike capability increased dramatically with the commitment by the US of five SSBNs to NATO. The SLBMs could be included in SACEUR's planning, although they would remain under 'complete American operational command and political control' (Osgood 1962: 4). This commitment was in response to a stated SHAPE need to provide a replacement for vulnerable, liquid fuel intermediate-range ballistic missiles (IRBM) which had been deployed in Italy and Turkey, as well as vulnerable strike bombers under SACEUR's command. (IRBMs were withdrawn by the US as part of the settlement to the Cuban Missile Crisis.) There was also concern about the increasing number of Soviet missiles in western Russia capable of reaching NATO Europe (Osgood 1964: 13).

Targeting for the Polaris assigned to SACEUR came within the scope of the SIOP, and was co-ordinated at SAC headquarters in Omaha. However, an onshore task force command, CFT-64, was established in Naples (headquarters of CINCSOUTH) for control of SSBN operations. According to CINCUSNAVEUR's 1963 report, 'target coverage' for SSBN assigned to SACEUR commenced on 1 April 1963 with USS *Sam Austin* (SSBN 609) on patrol in the Mediterranean. 'Since that first deployment,' the report continued, 'one of the three SSBN's assigned to SACEUR has been maintained on patrol in the Mediterranean.'[22]

The US Polaris fleet established a base at Holy Loch, Scotland, and by April 1962, six SSBNs were operating from that anchorage.[23] The fleet conducted regular patrols in the North Atlantic and Mediterranean. Some of these north Atlantic patrols were conducted under and for USCINCEUR, acting as the American Commander-in-Chief, while others were conducted in the Mediterranean under SACEUR. Thus, by April 1965, 111 patrols had been completed by SSBNs deployed in the Atlantic, of which 21 had been under USCINCEUR and 33 under SACEUR in the Mediterranean. Between July 1966 and June 1967 alone, 60 SSBN patrols were conducted under SACEUR in the Mediterranean with 69 in the Atlantic for USCINCEUR.[24]

With the deployment of Polaris SSBN, the US carrier force earmarked for NATO gradually withdrew from 'deep targets' and concentrated on 'regional plan' under SACEUR and especially under CINCSOUTH.[25] The USN, which had been developing smaller-yield atomic weapons for use on carrier-based aircraft, now emphasized the suitability of carrier-based airpower for the developing doctrines associated with tactical nuclear war. Thus in his 1964 report, CINCUSNAVEUR noted that his OPLAN 222–63 had been developed as guidance for the 'selected use of tactical nuclear weapons' in support of USCINCEUR's plans for using such weapons.[26]

The British had also developed an SSBN using the American Polaris SLBM. According to official statements, this force, eventually reaching 4 Resolution Class SSBNs carrying 16 SLBMs each, was dedicated to NATO, forming part of its strategic deterrent posture. Force targeting was included in SACEUR's General Strike Plan and co-ordinated with the US SIOP. However, the UK reserved the right to withhold its SLBMs for 'national' tasks in the event of a general nuclear exchange in Europe. Capable of hitting targets deep within the Soviet Union, the Royal Navy's SSBNs, held back in a continental exchange, would still provide a deterrent against Soviet strikes on England itself.[27] (In some ways this would be similar to Churchill's holding back of a part of the RAF fighter forces during the Battle of France for protection of England. If Western Europe fell quickly under a Soviet conventional and tactical nuclear assault, the UK would want to retain a force capable of deterring the Soviets.) The French did not deploy their Redoubtable Class SSBN until 1970, and it was not specifically earmarked for NATO tasks.

For the other members of the Alliance, all that was important was that the US had decided to earmark a portion of its SSBN force for NATO. It reinforced the link between the security of Europe and the American nuclear deterrent by designating strategic forces in European waters for SACEUR tasks.

It is important to note, therefore, that when, in 1961, President Kennedy announced that the US looked forward to 'the possibility of eventually establishing a NATO seaborne missile force which would be truly multilateral', the initial steps had already been taken to provide SACEUR with access to such a force. And, by the time the MLF idea had been abandoned, considerable sea-based strategic nuclear power had already been committed to the European command. As Robert Osgood noted at the time, the substance of the MLF (Multilateral Force) proposal had to do with control of nuclear weapons, which in the NATO context constituted a direct political problem and only indirectly a military one:

> For it is principally a problem of sustaining mutual confidence among allies, of accommodating vital interests and of satisfying demands of national equity and pride ... of achieving these political ends essential to the cohesion of the alliance, by adjusting the terms of military collaboration.
>
> (Osgood 1964: 1)

With regard to the sea-based use of nuclear weapons, there was collaboration to the extent that American forces were earmarked for NATO use in the event of war. The supreme commanders who would direct their use were Americans, but accountable for their use to the Joint Chiefs of Staff and ultimately to the President. In the planning and targeting of these forces, senior non-American allied commanders would also be involved.

In addition to providing more collaboration than this, the MLF was partially in response to the development of independent nuclear capabilities by the British and especially by the French, which might tend to undermine the American nuclear-use monopoly within NATO. Another problem was Germany. MLF was viewed as a means to bring the Germans into the nuclear club, by making NATO itself the fourth nuclear power.

This would be done by building 25 surface ships and placing 8 Polaris A-3 missiles, with a range of 2,500 miles, on each ship. The Europeans would build the ships and the US would supply the missiles. These ships would be assigned to SACEUR and would have multinational crews. The missiles would be under joint ownership and custody of the participating nations, unlike the tactical nuclear weapons then deployed in Europe. Neither the captain of each ship, whose nationality would vary, nor any of the national components of the crews, would be able to fire the missiles on their own. A NATO body would decide when to use them and an electronic signal would be beamed to each ship releasing the 'permissive link' only when this body decided.

The MLF idea reflected a growing preference for the basing of nuclear weapons at sea while at the same time providing partial solutions to the political problems of nuclear ownership, planning and decision. It was with regard to the decision-making procedures for firing the MLF missiles that the project appeared to flounder and was eventually discarded. Would unanimity amongst the MLF partners be required or, as the German Minister of Defense suggested, 'a system of weighted voting that would allow the use of the MLF even with some dissenting votes amongst the collaborators in the project' (Amme 1967: 163)? Would the controlling body be able to out-vote the United States? Would a veto on MLF use make a difference so long as the US maintained its Polaris and Minuteman capabilities?

Although the MLF never came into being,[28] some steps were taken to multilateralize nuclear planning in NATO. In December 1966, the Defense Planning Committee established the Nuclear Affairs Committee and the Nuclear Planning Group (NPG). The scope of these bodies would include sea-based nuclear weapons under British and mainly American control, but would not affect the exclusivity of national control.

The NPG established guidelines for the use of nuclear weapons in a document entitled 'Concepts for the Role of Theater Nuclear Strike Forces in ACE (Allied Command Europe)'. Based on these guidelines, SACEUR drew up his General Strike Plan (GSP) in the early 1970s. The GSP provided for two broad categories of nuclear warfare – selective use and general nuclear response. Selective employment would be used on a controlled or limited scale for demonstrative or tactical purposes, for example to contain a Soviet conventional breakthrough. Under general nuclear response the GSP provided for hitting a broad range of targets in conjunction with the execution of the SIOP. Targets to be struck were included in SACEUR's Scheduled Program, divided between his Priority Strike Program (PSP) and his Tactical Strike Program (TSP). Targets on both the PSP and TSP were co-ordinated or 'deconflicted' with the US SIOP target list.

The GSP was supported by forces in Europe kept on Quick Reaction Alert (QRA). At the normal peacetime level there were to be enough QRA forces to hit a certain percentage of the targets of SACEUR's PSP. Included in these forces are land-based aircraft, Pershing missiles and British and American sea-based SLBMs. At the advanced readiness level, more targets are covered by the addition of more land-based air forces, Pershings and SLBMs. At the maximum, additional land-based forces as well as sea-based air forces and more SLBMs are placed on QRA (US Senate 1973).

Most of the sea-based nuclear capability assigned to US forces earmarked for NATO was for projection purposes. But there was also, during these early years of the Cold War, an interest in nuclear anti-submarine weapons. American atomic tests in the late forties had raised the possibility that a nuclear charge underwater might be effectively used against submarines without harming nearby friendly surface forces. The 1951 MIT study, *Project Hartwell* (A Report on the Security of Overseas Transport), called for American deployment of nuclear ASW weapons as a means to ensure the safety of convoys and carrier task forces in a war with the Soviets. Small nuclear devices exploded at depths in excess of 1,000 feet would 'sink a submarine in a radius of over a mile while leaving surface vessels at a radius of a half of a mile unaffected'.[29]

The US Navy had already begun to develop depth weapon variants of aerial dropped bombs in the late 1940s. The first such weapon, an ASW type of the MK-7 bomb, entered the US nuclear stockpile in 1952, followed by smaller bombs in the late 1950s. A 1956 report by CINCLANTFLT noted that the US Atlantic fleet had achieved an 'atomic depth capacity'.[30] In support of the deployment of American nuclear ASW weapons, the US, sometimes in co-operation with the NATO infrastructure program, built storage facilities for warheads in the Atlantic and Mediterranean. By 1962, there were eight atomic underwater 'weapons shops' in the Atlantic region.[31]

In 1963, a new air-delivered nuclear ASW bomb, the B57, entered service. It could be dropped by land-based aircraft such as the P-3, and by carrier-based aircraft such as the S-3 and SH-3 helicopters. It would appear that provisions were made for release of B57s to allied navies under the program of co-operation agreements which governed possible use of American nuclear weapons by allied forces.

The first nuclear torpedo, the MK-45 ASTOR, also entered USN service in 1963. It could be launched from surface ships, submarines and aircraft and was said to be 'available' to NATO navies, although never deployed in their forces. Use of the ASTOR was discontinued because of difficulties of getting the firing platform far enough away to avoid being damaged by the blast.

Real advances in nuclear ASW came with the development of rocket assisted torpedoes which could take the nuclear charge far enough away. Two such systems were deployed. The ASROC (anti-submarine rocket) entered service in 1961 carrying a 1 kiloton warhead with a range of 6 nautical miles. It was installed on many American surface vessels. ASROCs carrying a conventional warhead were installed in a number of allied surface forces. In 1965, an inertially-guided submarine-launched range-controlled nuclear depth charge ASW rocket (SUBROC) was deployed. It carried a 1–5 kiloton warhead with a range of 20–35 nautical miles. Fired from a submarine's torpedo tubes, the SUBROC would break the surface of the water, travel through the air and re-enter the water to attack the target submarine (Cochran *et al*. 1984: 63, 267–71).

In time, the number of B57, ASROC and SUBROC nuclear ASW warheads would grow to over 2,000. Nuclear ASW could still not be used close to friendly forces nor in shallow waters near coastal areas because of effects of the blast. Thus, while the USN provided NATO with a nuclear ASW force of considerable magnitude, conventional ASW weapons still remained essential. Also to be considered was the escalatory effects of nuclear ASW use and therefore the tight political control on their use, even on the high seas.[32]

US WAR-AT-SEA PLANNING

The United States had committed its strategic and tactical nuclear forces to the NATO deterrent posture, threatening the Soviet Union with unspecified levels of atomic retaliation should it attack Western Europe at whatever level. By the late fifties, with the Soviet strategic nuclear capability growing, questions were raised as to the credibility of this deterrent for dissuading the Soviets either from limited aggression or even harassment of NATO, such as occurred during the Berlin Crisis. Deterrence would be more credible, and, in the event of a failure of deterrence, response much safer and controlled, if NATO availed itself of a wider range of response options.

As noted above, allied maritime strategy from the outset had been predicated upon making use of the seas to support SACEUR under a number of different contingencies. Seapower would play a role at each level of escalation up to, and including, full-scale strategic nuclear war in Europe.

In the early 1960s, the USN added to the flexibility of its planning by developing various 'war-at-sea' scenarios. The US experience during the Berlin Crisis of 1959 had prompted the then Chief of Naval Operations, Arleigh Burke, to initiate studies of the feasibility of responding to Soviet harassments on the ground in Europe, by applying naval pressure.[33] At sea, the US and its allies held marked superiority compared with the conventional imbalance on the ground in Europe. Paul H. Nitze, during his first tenure as Secretary of the Navy, initiated a comprehensive series of war-at-sea studies along the lines Burke had suggested. A response at sea to Soviet harassment might lead to limited war at sea which would be preferable to challenging the Soviets directly along the NATO frontiers because it carried less risk of escalation to nuclear war (Schratz 1980: 345).

The concept was very much in the classical Mahanian tradition. Mahan had tended to see the oceans as the no-man's land upon which the competing naval powers struggled in times of war for singular supremacy. Such supremacy, achieved through destruction of the enemy's capacity to wage war at sea, would foretell the resolution of the conflict in favor of the victorious naval power. A Soviet and NATO confrontation at sea would follow a similar pattern. As Lawrence Martin noted, the North Atlantic would be turned into a kind of Western Desert for a soldier's war which, 'one supposed, would be more likely to remain within the bounds of a regulated policy contest than one that wrought destruction on homelands and civilians even if only as the collateral effect of a would-be intra-military contest' (Martin 1965: 87).

The Nitze-initiated studies did embody this line of reasoning. To quote from the introduction to Volume I of the study:

> The operations here analyzed constitute an exclusively maritime war, 'war restricted to the sea', where the objective is to coerce the antagonist solely by maritime pressures. Thus nearly all combat takes place at sea. Land, overland airspace, and inland waters are, in general, taken as sanctuaries.
>
> War restricted to the sea is a special case of 'war at sea' defined literally as the war, or that part of a wider war, which takes place on, in and over the sea, where the war as a whole may or may not be exclusively, or predominantly maritime.[34]
>
> ... combat actions are confined to the maritime environment. There is no concurrent land combat. The political objectives of the contes-

tants are considered to be reasonably well defined and do not encroach upon the vital interests of any nuclear power. This is recognized by all of the major contestants.[35]

What the study called the 'governing issue' of its analysis of war restricted to the sea was the 'capacity of the US side to operate its combat and merchant sea forces as needed', in other words, to make straight use of the sea. In this sense the criteria for the success of a war waged only at sea were similar to those used in judging the success of a sea campaign fought in conjunction with land campaigns. These included: damage and loss rates for both sides; loss of merchant and fishing ships; economic value of losses from disruption of sea transportation; and 'changes in military posture, i.e. alterations to strengths, situations or deployments so as to alter the options available to either side'.[36]

In a general NATO war success of the allied navies would be judged by similar criteria, the 'governing issue' being the extent to which the strategic value of the seas could be used to support the overall allied effort. Thus, the actual implementation of a war-at-sea approach would not differ substantially from what the NATO maritime forces were likely to do in the event of a more general conflict, with the important exception that in such a war there would be little or no projection of power ashore. NATO would win by denying to the Soviets the use of the sea and preventing the USSR from closing off sea-based options to NATO. A war restricted exclusively to the sea might well then be the first real battleground for NATO and Warsaw Pact forces – a battleground selected by both sides, according to the theories, because of the reduced chances of escalation and the enhanced prospects of political control. This would be so even if the sea war involved the use of tactical nuclear weapons against targets at sea, i. e. employment of atomic ASW weaponry.

On the basis of the existing sources, it is impossible to say how widely, or how seriously, war-at-sea scenarios were taken amongst naval leaders in the allied navies. One British analyst has suggested that, while discussed, such scenarios received little attention from the Royal Navy. On the other hand, former Admiral of the Fleet and Head of the NATO Military Committee, Lord Hill-Norton, did not discount the possibility of an initial NATO and Warsaw Pact confrontation being confined to the sea. He pointed to the example of 1939 and 1940, when the Western Front was quiet while action took place at sea.[37]

What can be said for certain is that beginning in the early 1960s, war-at-sea plans are mentioned in the annual reports of the American Commander-in-Chief Atlantic Fleet who was also SACLANT, and who had under his direct command those forces capable of implementing a war-at-sea approach. Thus in CINCLANTFLT's report for 1962, mention is made of OPLAN 115–61 which provided for 'reciprocal measures instituted at sea against the USSR, GDR or designated Soviet Bloc shipping in varying degrees of severity'.[38] In November 1961, the USN deployed an anti-submarine task group to the northeast Atlantic 'for possible employment in the harassing of Soviet naval operations as a counter to Soviet harassment in Berlin'.[39]

By 1967 CINCLANTFLT had developed OPLAN 2300 entitled 'War At Sea Plan'. It provided for fleet operations in contingencies involving the Soviet Union. Such operations were to be 'specifically designed to apply a measure of force in support of a national policy of determined persuasion, or alternatively, may be implemented as a controlled response to Soviet military aggression short of strategic nuclear attack'. The plan outlined a variety of naval actions to be employed in situations 'wherein the US desires to employ military force against the Soviet Union under circumstances in which implementation of the general war plan is neither a rational nor credible threat'. This was to be the 'measured' application of military force, with increasing escalation towards the 'highest intensity of operations' which would entail 'the entire Atlantic fleet in a complete confrontation with Soviet forces at sea in warfare short of the strategic use of nuclear weapons'. The actions taken by the US fleet would be controlled by 'rules of conduct and engagement reflecting national policy objectives'.[40]

The war-at-sea planning within the USN seemed to have been the result of the more general trend in American strategic thought which stressed the feasibility and desirability of flexible response, the measured application of military force even in response to a Soviet attack. By shifting the application of force to the seas such an approach played upon the strategic value of the seas to both sides. If sufficient losses could be inflicted upon Soviet maritime forces, including merchant ships, then Moscow might be persuaded to negotiate. Implicit in war-at-sea planning, however, was the assumption that with its superior naval forces the US would prevail at sea. In this sense, such planning sought to take advantage of the secondary strategic value of the seas to the Soviets. If Russia did not depend heavily on the seas to support either its economy or its military power in central Europe, then striking at Soviet vessels would not elicit a massive retaliation by the Soviets.

For the US and NATO a loss at sea would be much more serious. Disruption of the ocean sea lanes could well trigger allied escalatory steps because it would mean the isolation of allied forces facing the Soviets in central Europe and along the northern and southern flanks. The dependence of NATO upon securing use of the seas was certainly appreciated in Moscow. This, in turn, imposed limitations on the effectiveness of purely naval action in persuading the Soviets to negotiate. Rather than await the outcome of a naval confrontation which, when won by NATO, would improve the overall alliance posture, the Soviets would more than likely move on the ground or take some other escalatory step that could undermine the political and strategic utility of a NATO naval victory, i.e. striking at port facilities with nuclear weapons, or moving conventional forces quickly to take control of key NATO ports.

Problematic considerations such as these beset the general concept of flexible or graduated response. They limited, but did not entirely invalidate, US plans to come to the aid of Europe by first waging war against the Soviets at sea. Such planning only serves to emphasize again the relevance of seapower to the full spectrum of the Alliance's deterrent posture.

THE COLD WAR AT SEA: NATO MARITIME STRATEGY IN THE EARLY YEARS

With the establishment of its maritime command structures and the development and earmarking of nuclear and conventional maritime forces to those commands, NATO had quickly become a maritime alliance. Seapower was a part of the allied deterrent posture. Given the emphasis during the Cold War upon nuclear deterrence, it is not surprising that the allied maritime posture included forces capable of projecting nuclear force ashore. The reorientation of the US Sixth Fleet away from conventional support of the land battle to a nuclear projection role, and the assignment of SSBNs to SACEUR, were consistent with NATO's reliance upon the strategic nuclear capabilities of the United States.

Yet, it is clear from the very nature of the command structure, from the types of maritime forces the allies were building and from the exercises that took place, that NATO maritime strategy during the Cold War encompassed much more than providing sea-based nuclear projection forces. In his 1958 report to the Secretary of the Navy, the US Chief of Naval Operations noted that:

> 146 [ships] in the Atlantic fleet and the entire Sixth Fleet, participated in a series of the most extensive NATO exercises to date.... These were directed by SACLANT and SACEUR and embodied every type of naval operation. Also in the group were ASW mining and shipping

control exercises. The area of operations included the Norwegian Sea, the Eastern Atlantic and the Mediterranean... a nuclear powered submarine, FORRESTAL Class carriers, guided missile cruisers and guided missile submarines were involved ... all of which were geared to atomic warfare.[41]

It was the emphasis NATO maritime planning placed upon providing for traditional sea securing operations, intended to allow for reinforcement and resupply of conventional land and air forces, that drew much of the criticism directed at the Alliance's maritime strategy during the Cold War. Writers such as Lawrence Martin and Sir Peter Gretton viewed this emphasis as being anachronistic and 'out of tune' with the overall allied strategy which stressed nuclear deterrence.[42] Since there was little prospect of a prolonged conflict on the continent and the allies were turning their backs upon the goals of increased ground and air forces proposed in the early 1950s and suggested again by the Kennedy administration, there seemed little reason to maintain the level of forces dedicated to protecting the sea lanes of communication, continually advocated by allied naval leaders.

Other writers, such as Jurgen Rohwer, have argued that NATO overreacted to the Soviet submarine threat in the 1950s. The Soviet Union was not building forces capable of cutting the allied sea lanes of communication across the Atlantic, but rather was developing a coastal submarine defense force whose primary purpose was the protection of the Soviet Union from American carriers and amphibious forces (Rohwer 1975). Given the unlikelihood of a prolonged conventional war and the defensive posture of the Soviet submarine force, others have suggested that the allied maritime effort during the Cold War was the result of bureaucratic politics and inter-service rivalries. Western Admirals, looking to justify larger fleets, pressed for more general purpose maritime forces beyond nuclear projecting forces. Thus they spoke in terms of a third battle of the Atlantic.[43]

While bureaucratic politics and inter-service rivalries no doubt had something to do with the development of a wide-ranging allied maritime posture, a strong case can be made for the strategic soundness of this posture during the Cold War and the early days of NATO. The strength of this case, moreover, does not rest entirely upon the fact that the 'flexible response' emphasis of the allied maritime forces proved, in the long run, to be the most consistent with overall allied needs. A number of other considerations have to be taken into account.

In the first place, it is not entirely accurate to juxtapose an allegedly anachronistic NATO maritime posture and strategy with a supposedly well defined and consistent overall allied strategy. There is little doubt that the Alliance was publicly emphasizing nuclear deterrence during its early years. And the Lisbon goals for conventional forces were never met, placing in doubt NATO's overall conventional war-fighting ability. Yet, there would be no certainty as to what the exact allied response would be in the event that deterrence failed. In particular, considerable doubt existed (as it still does) over the use of nuclear weapons, strategic and theater/tactical. General Robert C. Richardson III, USAF, one of those who worked on NATO's early nuclear strategy, has pointed out that the 'new approach' was never wholly implemented in terms of the force posture of the Alliance, although the allied governments gave it their approval. 'No one could argue with assurance,' recalls Richardson, 'that a NATO strategy of primary reliance on nuclear weapons would actually work. The needed stockpile of tactical nuclear weapons was not yet available, nor had a requested force posture been designed' (Richardson 1981: 36). The NATO governments had not allowed for the two elements that would make early use of nuclear weapons 'operationally viable', namely, dispersal of nuclear weapons in the standing forces and configuration of the forces to fight a tactical nuclear war.

Second, even if the Alliance was assured of being able to execute its atomic strike plans, those early plans called for conventional forces to prevent a rapid Soviet break-through into Western Europe. Such forces would be required in spite of the fact that the strategic air offensive against the Soviet Union itself was expected to 'virtually eliminate the Soviet Bloc industrial capacity' at the outset of the war. With surviving stockpiles, the Soviets might not have experienced serious shortages until D+4 to D+7 months. Thus, as one study put it: 'The outcome of the ground battle is contingent upon the outcome of the air battles and upon the ability of Allied ground forces to make the Soviet ground forces concentrate and present good targets for attack with atomic weapons'(as quoted in Rosenberg 1981/2: 37).

Third, given the uncertainties over nuclear weapons use, and the continuing build-up of Soviet nuclear forces in the 1950s, heavy reliance by NATO upon nuclear deterrence was subject to question almost simultaneously with the advent of the 'new look' strategy. As early as 1957, SACEUR General Lauris Norstad argued that NATO's forces in Europe required a conventional capability to engage in resistance against 'limited or local attacks'. He rejected the idea that they should serve as a trip-wire or plate glass, meant only to touch off the activation of nuclear forces.[44] By the late fifties, the Alliance's leadership began speaking in terms of a 'pause', a definite firebreak between an initial, conventional attack by the Soviets and nuclear retaliation by NATO. Conventional forces would bring about such a pause. And then, with the coming to power of the Kennedy administration and Secretary of Defense Robert S. McNamara, the US in the early 1960s began to stress conventional deterrence, and the need to rely less upon strategic nuclear weapons.

Fourth, as long as NATO could not rule out some form of conventional war-fighting ability, allied planning to secure use of the seas for the transportation of reinforcements could not be said to be at variance with the overall NATO posture. Plans which called for the immediate execution of an atomic offensive rested upon assurance of adequate conventional forces. But, as the 1955 Evaluation of the Atomic Offensive cited above pointed out: 'There is no assurance that present allied ground forces, even with air superiority in their favor, are adequate to fulfill the conditions for successful use of atomic weapons on Soviet ground forces' (as quoted in Rosenberg 1981/2: 36). Thus there was some anticipation that sea-lifted reinforcement and resupply would have to take place. With nuclear strikes against Soviet naval facilities by the USN and wide use of conventional ASW, there was a general confidence in NATO being able to secure the sea lanes of communication. However, plans had to be drawn up and forces earmarked to achieve these objectives.

A fifth consideration is that in planning to secure the seas, allied military leaders were fully aware of the limited ranges of the USSR's major sea denial forces, especially its submarines. The Soviets may indeed have built and maintained their submarine force for defensive purposes, but this was no assurance that it could not be used to contest control of the immediate coastal waters vital to NATO. Although allied naval leaders did speak publicly about a third battle of the Atlantic, they anticipated not a fight on the high seas, but a struggle in the waters approaching the continent. This thinking was reflected in a memorandum prepared for the US Secretary of Defense by General Omar Bradley in 1951. He stated that no major problems were expected in the Western and mid-Atlantic, rather:

> The northeastern Atlantic and Mediterranean are forward areas in which the level of hostile action would be relatively high and in which the active support of operations on the continent is extremely important ... [there should be] continuity of action ... [in the ASW campaign] and ... maximum flexibility for control and routing of convoys. Command arrangements in the Atlantic and also the Mediterranean must

satisfy the requirements for active support of the Supreme Allied Commander Europe.[45]

Finally it must be appreciated that whatever the announced strategic doctrine, the expectation was that a war with Soviets in Europe would evolve out of a crisis, a period of rising tensions. During this period, steps would be taken to secure the seas. While some of these steps would be in preparation for the execution of sea-based nuclear strikes, others would be carried out regardless of specific expectation as to the nature and length of an impending war. There would have to be heightened surveillance of Soviet submarine movements, utilizing the ocean floor SOSUS system, and mobilized surface, subsurface and airborne maritime forces. For example, at the time of the Suez Crisis, the US Navy established a submarine anti-submarine barrier along the Greenland–Iceland–United Kingdom gap. Initially seven submarines, supported by aircraft, were assigned to patrol between 17 and 20 November 1956. Four submarines remained on station until 13 January 1957.[46] During the Berlin Crisis of 1961, the US Navy deployed an anti-submarine carrier group to northern Europe.[47]

These crisis deployments were only of American forces, since the situations were not such that the allied maritime commands had been fully activated. Given its commitment to NATO, and to its own forces already on the continent, the United States would necessarily take the lead in mobilizing maritime forces during a crisis. Throughout the Cold War the annual report of US CINCLANTFLT (who was also SACLANT), pointed out that American plans, in addition to being co-ordinated with NATO plans, provided for the performance of national tasks in the event that: '(a) SACLANT is delayed in assuming control; (b) NATO fails to function; (c) the US should decide to take unilateral action.'[48] In his 1955 report, the second SACLANT, Jerauld Wright, noted that CINCLANTFLT's plans are prepared in such a way that 'initial deployments and operations are the same in almost all cases whether forces remain under US commands or are transferred to NATO commanders'.[49]

When fully deployed and placed under allied commanders, the collective maritime forces of the Alliance would, in the event of war, have tried to secure and exploit the seas. Soviet sea denial forces could have mounted significant challenges, particularly in the coastal waters within range of land-based naval aviation. Throughout the Cold War, the Soviet Union's maritime forces were increasing their capabilities. Nevertheless, as late as 1968, a lengthy assessment done by the SACLANT staff concluded that the Alliance could expect to secure control of the seas against Soviet forces in a major conventional maritime struggle. The struggle could take over a month, and there would be significant losses on both sides, but in the end, NATO would prevail at sea.[50]

It is evident, therefore, that the wide-ranging maritime strategy adopted by the Alliance during the Cold War, and the various national forces developed to support that 'flexible' strategy, afforded NATO a maritime superiority that extended well beyond the ability to project nuclear force ashore. Given the almost inherent uncertainties that attended the allied military and political doctrines, this effort to perpetuate Western seapower, in all its forms, was by no means inconsistent with the overall NATO strategic posture.

At the same time, it cannot be denied that, in the presence of overwhelming NATO nuclear superiority, the relative importance of superior seapower, save its nuclear elements, was circumscribed. As long as allied leaders felt confident that they could deter a Soviet conventional assault on Western Europe by threatening nuclear retaliation, however unspecified was the threat, seapower could not be considered a decisive element in the general Euro-centric balance of power. NATO was unwilling to do without the ability to secure, deny and exploit the seas, but this ability only represented an additional and not crucial margin of safety to the allied deterrent posture.

As the nuclear balance began to shift in favor of the Soviet Union, and as the conventional land and air balance in Europe continued to favor the Warsaw Pact, the relative importance of NATO's seapower would increase. With the allied posture moving unevenly and haltingly toward flexible response, the flexible posture adopted by NATO's maritime forces in the Cold War would come to be looked upon as a necessity. Yet, by then, the maritime superiority of those early years appeared to be diminishing. NATO was becoming more of a maritime alliance, but one which could not confidently expect to support the exercise of military force ashore from a position of absolute control of the sea.

Chapter 4
Soviet maritime forces and flexible response

SOVIET MARITIME DEVELOPMENTS

The statistical tables 4.1 to 4.3 present a summary of Soviet and Warsaw Pact naval forces, as well as American and allied forces by the mid-1970s.

In absolute numbers, the Soviet and Warsaw Pact navies, as well as the navies of the Western allies, declined during the postwar era. As ships built during the Second World War became obsolete, they were scrapped and replacements were seldom on a one-for-one basis. The newer ships, and naval aviation, tended to have more unit firepower than their predecessors. By the mid-1970s, NATO still held a substantial lead over the Soviets in the number of naval forces. While all commentators (and most professional sailors) stress that the naval balance cannot be measured by numbers alone, it is evident that although the Soviet Navy expanded, the Western navies were not standing still. For example, between 1958 and 1971, the Soviets took delivery of 76 destroyer-escorts at rates of 4–7 new ships per year. During the same period, the NATO nations took delivery of 224 destroyer-escorts, with 53 going to the US Navy and 47 to the Royal Navy. In the category of destroyer-escorts armed with anti-ship missiles (SSM) and anti-aircraft missiles (SAM), the Soviets took delivery of 31 new ships between 1958 and 1971, while NATO acquired 79 (the USN accounting for 24). During these same years, the Soviets dramatically improved their nuclear submarine capability, taking delivery of 103 vessels including SSBN, nuclear attack submarines (SSN) and nuclear cruise missile attack submarines (SSGN). But NATO nations deployed well over 100 nuclear submarines including the Polaris and later Poseidon SSBN (MccGwire 1973b: 146).

In general, as Michael MccGwire notes, 'over the fourteen years spanning 1958–1971, the West had taken delivery of two to three times the number of major combatants [aircraft carriers, cruises, destroyer-frigates, escorts and submarines] as had the Soviet Navy'. He adds that 'if

Soviet maritime forces and flexible response   77

Table 4.1 Soviet Navy 1974

SHIPS – MAJOR COMBATANTS	SOVIET 1974 BY FLEET					UNITED STATES 1974
CARRIERS	NORTH	BALTIC	BLACK	PACIFIC	TOTAL	
ATTACK CARRIERS (CVA)	–	–	–	–	–	13
ATTACK CARRIERS, NUCLEAR (CVAN)	–	–	–	–	–	1
HELICOPTER/STOL (CVSG)	–	–	–	–	1	–
HELICOPTER CARRIER (CVH)	–	–	2	–	2	7
SUBMARINES						
BALLISTIC MISSILES (SSB)	15	–	–	8	23	–
BALLISTIC MISSILES, NUCLEAR (SSBN)	39	–	–	12	51	41
ATTACK CRUISE MISSILE (SSG)	13	3	5	7	28	–
ATTACK CRUISE, NUCLEAR (SSGN)	28	–	–	14	42	–
ATTACK TORPEDO (SS)	36	40	27	28	151	20
ATTACK TORPEDO, NUCLEAR (SSN)	25	–	–	10	35	58
CRUISERS						
LIGHT CRUISERS, SAM & SSM (CLGM)	6	2	5	3	16	–
LIGHT CRUISERS, SAM, COMMANDSHIP (CLG/CC)	–	–	1	1	2	8 (NO SSM)
LIGHT CRUISER (CL)	1	4	4	2	11	1
DESTROYERS & ESCORTS						
DESTROYER (DD)	8	6	8	14	36	94DD (SOME WITH SAM)
DESTROYER, SAM (DDG)	5	2	3	2	12	
DESTROYER LEADER, SAM (DLG)	1	4	10	4	19	
DESTROYER, SSM (DDGS)	–	–	3	1	4	
DESTROYER, SAM & SSM (DDGSP) POINT DEFENSE	5	2	2	–	9	
PATROL ESCORT (PCE) (SOMETIMES LISTED AS ASDF, DESTROYER ESCORT) (CLASSES: MIRKA, PETYA, RIGA, KOLA)	20	23	35	24	102	66 GENERAL
PATROL ESCORT, SAM (PCEP)	3	6	7	4	20	6 (SSM, SAM)
MAJOR COMBATANTS					564	379
SMALLER VESSELS						
PATROL ASW (PCS)	35	90	35	45	205	130 (COAST CUTTER)

Table 4.1 Continued

	SOVIET 1974 BY FLEET					UNITED STATES 1974
	NORTH	BALTIC	BLACK	PACIFIC	TOTAL	
PATROL GUNBOAT, SAM & SSM (PGGP)	–	4	4	–	8	
PATROL TORPEDO BOAT (PTF)	20	70	25	45	160	17
PATROL, ANTI-AIRCRAFT & SSM (PTFG)	25	35	25	35	120	2 WITH MISSILES
PATROL BOAT, SSM (PTG)						
AMPHIBIOUS LANDING SHIPS AND CRAFT	38	74	63	74	249	APPROX. 150
MINESWEEPERS						
COASTAL (MSC)	15	35	25	20	95	–
FLEET (MFC)	45	55	40	45	185	31
OCEAN-GOING SUPPORT						
OILERS, SUBMARINE TENDERS, WATER CARRIERS, INTELLIGENCE, AMMUNITION, HYDROGRAPHIC RESEARCH	93	54	55	87	279	156
NAVAL AVIATION (EXCLUDES LAND-BASED TACTICAL AIR)						
LONG-RANGE, ANTI-SHIP MISSILE (ASM)	150	75	40	35	300	750 FIGHTERS 1,270 ATTACK CARRIER BASED
ASW	135	95	50	30	310	400 LRPA

Source: *Jane's* 1975; IISS 1975; Berman 1975

Table 4.2 NATO maritime forces 1975

REGION AND COUNTRY	CARRIERS CVA(N)	CVH	CRUISERS	DESTROYERS FRIGATES ESCORTS CORVETTES	SUBMARINES SSBN	SSN	SS	NAVAL AVIATION CARRIER	LAND	SMALL PATROL COMBATANTS	MINESWEEPERS LAYERS AND AMPHIBIOUS
ATLANTIC											
UNITED KINGDOM	1	2	2	64	–	4	8	23	58	–	11	42+7 amphibious
CANADA	–	–	–	15	–	–	–	3	–	25	–	–
PORTUGAL	–	–	–	29	–	–	–	4	–	12	36	16
US 2ND FLEET	4	2	8	68	31	21	5	360	180	–	20 amphibious
NORTHERN FLANK											
NORWAY	–	–	–	7	–	–	–	15*	–	9	46	14
DENMARK	–	–	–	10	–	–	–	6	–	–	51	19
WEST GERMANY	–	–	–	23	–	–	–	30	–	20	10, SSM	70
NETHERLANDS	–	–	1	35	–	–	–	6	–	24	–	53
BELGIUM	–	–	–	–	–	–	–	–	–	–	–	28
MEDITERRANEAN											
ITALY	–	–	3	32	–	–	–	9	–	38	18 some with SSM	57+2 amphibious
GREECE	–	–	–	15	–	–	–	7	–	12	19	23+8 amphibious
TURKEY	–	–	–	21	–	–	–	13	–	14	60	29+2 amphibious
US 6TH FLEET	2	1	5	13–15	3–5	12	–	180	60	–	15 amphibious
FRANCE	2	–	2, SSM	45**	3	–	20	–	110	30	27	75+2 amphibious

*Norway has only coastal submarines.
**French and some other allied navies were deploying Exocet SSM.
Source: *Jane's* 1975; IISS 1975

Table 4.3 Major naval ship construction, Warsaw Pact and NATO 1955–74

Type/Class		1955–64	1965	1966	1967	1968	1969	1970	1971	1972	1973	1974	Ten Years 1965–74	Five Years 1970–4	In Service 1975[a]
SUBMARINES															
Ballistic missile, nuclear:	WP	9	3	–	1	4	5	5	7	8	8	8	46	36	55
	NATO	29	3	8	2	2	1	–	1	–	1	1	19	3	48
Ballistic missile, diesel:	WP	29	–	–	–	–	–	–	–	–	–	–	–	–	20
Cruise missile, nuclear:	WP	10	5	6	5	3	2	1	3	2	2	2	26	10	41
	NATO	1[b]	–	–	–	–	–	–	–	–	–	–	–	–	–
Cruise missile, diesel:	WP	26	2	2	2	1	–	–	–	–	–	–	7	–	25
	NATO	1[b]	–	–	–	–	–	–	–	–	–	–	–	–	–
Attack, nuclear:	WP	26	–	–	1	3	3	4	3	3	3	3	23	16	34
	NATO	21	–	3	8	5	10	4	9	4	3	4	16	24	72
Attack, diesel:	WP	127	6	6	6	1	2	1	–	–	1	2	25	4	163
	NATO	59	9	11	6	6	5	3	1	4	15	4	64	27	141
SURFACE VESSELS															
Attack carriers:	NATO	10	1	–	1	–	–	–	–	–	–	–	3	1	18
Cruisers:	WP	1	–	–	1	–	1	–	–	1	1	1	5	3	18
(9,500 tons +)	NATO	8	–	–	–	–	–	–	–	–	–	2	2	2	12
Large ASW ships:	WP	18	–	–	3	4	4	2	3	5	3	2	26	15	42
(4–9,500 tons)	NATO	64	2	6	5	1	6	16	11	13	7	7	74	54	143
Destroyers:	WP	21	–	–	–	–	–	1	1	2	2	2	8	8	71
(2–4,000 tons)	NATO	90	8	9	9	9	5	2	2	2	2	2	50	10	242
Escorts:	WP	49	8	8	13	8	8	–	–	–	–	–	45	–	110
(1–2,000 tons)	NATO	37	1	7	3	1	–	4	2	2	1	1	22	10	59
Patrol vessels:	WP	178	11	8	11	10	8	5	6	8	8	8	83	35	217
(250–1,000 tons)	NATO	27	1	–	2	4	2	2	2	4	6	6	29	20	102

[a] Differences between construction totals and vessels in service are caused by transfers, retirements and losses.
[b] Experimental only.
Source: IISS 1975: 88

account were taken of relative size and combat capability, the disparity was more like four to one' (MccGwire 1975: 531).

At the same time, it was possible to exaggerate the sheer numerical increases on the part of the Soviets. The greatest increase in numbers for the Soviet Navy was in the realm of minor combatants (usually patrol boats of 100 to 1,000 tons). Between 1966 and 1976, the Soviets took delivery of 480 minor combatants, while the USN acquired only 41 in this category. And while the number of US major combatants declined between 1966 and 1976, the USN taking delivery of 101 such vessels, the Soviet Union acquiring 131, for the largest ships, over 10,000 tons, the US led the Soviets by 30 to 3. In the category of replenishment ships (essential for extended voyages and sustained ocean combat) the USN took delivery of 23 to the Soviets' 4 between 1966 and 1976. Thus, US Department of Defense comparisons of 249 'combatant' deliveries to 766 for the Soviets must be qualified by the large numbers of smaller ships (Nitze *et al.* 1979: 228). (Although, as is noted below, the smaller Soviet and Warsaw Pact vessels, when armed with SSM, could constitute a serious threat particularly in the Baltic and coastal waters.)

The Soviets followed the American lead in the development of sea-based intercontinental missiles using both diesel and nuclear powered ships (SSB, SSBN). Between 1965 and 1974, the Soviets built 46 SSBNs to NATO's 19, 36 of these during the five-year period from 1970–5, during which NATO built only 3 SSBNs (IISS 1976: 88). The range of the initial Soviet SLBMs was shorter than that of the US Navy's Polaris. Even so, by 1967, the Soviets could hit all the major cities of NATO Europe from submarines firing from Soviet coastal waters (Hezlet 1967: 253). The new Delta Class SSBNs, which came into service in the early 1970s, with their 12 SS-N-8 SLBMs, afforded the Soviets 'the capability to cover virtually the whole of the NATO land mass from the northern Norwegian or Barents sea'.[1] This included parts of North America. As the Soviets' SSBN capability approached that of the United States and Britain, older, mainly diesel, missile-firing submarines remained in service, providing what some analysts termed 'frontal artillery' for Soviet land forces in Europe (Nitze *et al.* 1979: 398). By 1975, the Soviet Navy's SSBN force exceeded the combined NATO total by 55 to 48, although the USN had also improved its capability with the deployment of the multiple warhead Poseidon SLBM.

It was not, however, the development of capabilities similar to those of the NATO navies which concerned allied naval leaders, but rather the Soviet emphasis on weapons and weapons' platforms which appeared to be especially developed to counter the kind of forces NATO would rely upon. Scenarios of future confrontations at sea did not pit like-vessels against each other. The Soviets, for example, had not built modern attack carriers to counter the US carrier-based projection capabilities. 'With very few exceptions,' noted US Navy Chief of Naval Operations Elmo Zumwalt in 1972, 'US ships are not designed to fight Soviet ships of similar classes.'[2]

To counter NATO surface forces (those forces which would be dedicated to the tasks of reinforcement and resupply and projection) the Soviets had appeared to emphasize anti-ship missiles. In the early 1960s they began placing the Shaddcok surface-to-suface missile (SS-N-3), with an optimum range of 300 nautical miles (n.m.) on submarines and cruisers. By the late 1960s and early 1970s, newer generations of SSM were being deployed on submarines and surface combatants, including smaller vessels like the Nanuchka Class patrol gunboat. This small ship carried the SS-N-9 with a range of 60 n.m. (*Jane's* 1974: 634; McGruther 1978: 98). It was a Soviet SSM, fired from a smaller vessel, which sunk the Israeli destroyer *Eilat* in 1967. In the European coastal waters, the Warsaw Pact had nearly 1,800 smaller craft, many with SSM, while NATO deployed less than 600.[3] The US Navy's view was that 'these small craft would play an important role in conflicts around the periphery of Europe where they would be used to assist in cutting our sea lines of communication to our principal sources of supply and to our allies'.[4]

The Soviets had also been placing air-to-surface missiles on their long-range naval aviation. Badger bombers, with ranges of over 2,000 n.m., first carried Kipper ASM (As-2) then Kelt (As-5) with missile ranges of 100 n.m. (*Jane's* 1974: 636). Unlike the NATO allies, the Soviets had developed both nuclear and diesel powered cruise missile submarines (SSG and SSGN), capable of firing an anti-ship missile 30 n.m. Between 1955 and 1974, the Soviets had built nearly 70 of these ships, 36 between 1970 and 1974 alone. By 1975 they had 66 in service (41 SSGN, 25 SSG, while NATO had none (IISS 1976: 88). The Soviet Navy's anti-ship missile capability which, by the early 1970s, was based upon over 600 missile-firing surface ships, aircraft and submarines (Zumwalt 1976: 520), in addition to large numbers of diesel and nuclear torpedo-firing attack submarines (SS and SSN), presented NATO with a three-dimensional threat to its surface forces.

Allied naval leaders became especially concerned over the potential of Soviet sea denial forces as a result of the events which took place in the eastern Mediterranean during the Jordanian crisis of 1970 and the Middle East War of October 1973. In 1970, the Soviet Navy had interspersed missile-firing surface ships amongst the ships of the US Sixth Fleet (Kidd

1972: 26). During the 1973 War, the number of Soviet ships in the Mediterranean increased from an average of 35–40, to 95, including 25 submarines (Wilson 1976: 13). As Dov Zakheim has noted, the Soviet Navy's primary role was sea denial 'or at least neutralization of US forces, so as to ensure the continuation of the Soviet airlift' (Zakheim 1978a: 109). A similar interspersion during a NATO crisis raised the possibility of a 'D-day shoot-out' at sea when a war began. Allied naval forces would be subjected to concentrated air, surface and subsurface attack, raising the further prospect of a long and possibly indecisive campaign for control of the seas (Wilson 1976: 13).

By the early 1970s, Soviet naval exercises seemed to give credence to Western apprehensions. In Exercise OKEAN 70, simulated carrier task forces moving through the Greenland–Iceland–United Kingdom gap (GIUK), were attacked by 10 missile-armed surface combatants, 30 submarines and subjected to 400 sorties of aircraft arriving in waves (Daniel 1978: 223, 225). Soviet naval aviation took advantage of Soviet facilities in Cuba. Following the OKEAN 70 exercise, two Bear bombers flew from bases near the Kola Peninsula over Soviet ships operating in the North Sea on to Cuba. These aircraft would 'provide eyes for the approximately 300 Badgers armed with Kepper and Kelt stand-off ASMs' (Kipp 1977: 213).

Other maneuvers appeared to simulate a period of rising East–West tension with long-range reconnaissance as well as ASW searches beyond the GIUK gap. Submarine visits to Cuba increased, including visits by SSBN to Cienfuegos which, in the autumn of 1970, brought protests from the United States. The presence of Soviet missiles in Cuba was regarded as a violation of the 1962 agreements ending the Cuban Missile Crisis. In European waters, especially in the Baltic, Soviet and other Warsaw Pact forces were demonstrating a growing amphibious capability (Daniel 1978: 225).

In 1975, the Soviets conducted an exercise which involved naval activity on a global scale. OKEAN 75 saw over 200 ships from all fleets, as well as large numbers of naval aircraft (flying over 700 sorties), engaged in maneuvers in the Atlantic, Pacific and Indian Oceans. It appeared that the Soviets were simulating a growing East–West crisis, as they first deployed long-range naval aviation to Cuba, Yemen, Somalia and Africa. Subsequent movements appeared to test capabilities for interdiction of the sea lines of communication (SLOC), particularly in the North Atlantic. As one analysis suggested, OKEAN 75 'showed planning for early attrition of hostile forces on the high seas before potentially hostile forces can reach the primary Soviet defense zone' (Watson and Walton 1976: 94).

Despite their extended scope, the OKEAN exercises were less extensive in terms of sea area covered and numbers of ships and aircraft involved than the major NATO exercises. The latter could see upwards of 300 ships, including several strike carrier task forces, at sea at once. However, the extent of Soviet naval maneuvers by the early 1970s did demonstrate a capability for deployment outside the coastal waters and shallow seas in which previous exercises had taken place. Moreover, Soviet naval forces enjoyed the benefit of being under a single national command at all times, whereas those which would be employed by NATO remained under the national command of several governments and were earmarked for allied use only in time of crisis or war. In the Atlantic and Mediterranean, the Soviets were establishing port facilities which aided in the forward peacetime deployment of Soviet vessels outside European coastal waters. By the mid-1970s, up to five SSBN and attack submarines could be found patrolling the North Atlantic, along with a couple of surface combatants. Intelligence, research and replenishment ships could also be found in the Atlantic (Weinland 1975: 411). To these must be added fishing and merchant vessels to which NATO ascribes a military potential, at least in terms of information gathering.

In the Mediterranean, the peacetime Soviet presence had grown considerably since the early 1960s with the involvement of the USSR in the continuing Arab–Israeli turmoil. With the establishment of port facilities in Egypt and Syria, the Soviet Navy was in a better position to sustain the presence along NATO's southern flank. By 1973, there were between 35 and 40 Soviet ships normally in the Mediterranean, including up to 10 attack submarines, 3 to 4 cruisers, 9 to 12 destroyers, 1 to 3 amphibious ships, along with various support ships (Wilson 1976: 13).

While the meaning and purpose of the enhancement of Soviet seapower seemed unambiguous to US and allied naval leaders,[5] the strategic literature on the Soviet Navy contains significant areas of dispute as to the rationale for this development and of the likely wartime employment of the Soviet naval forces.[6] In general, the debate was between those who saw Soviet naval developments as essentially defensive and those who believed that the Soviet Navy had gone beyond its primarily defensive purposes and was now embarked upon, and increasingly capable of, achieving global military and political objectives.

Michael MccGwire has been foremost amongst those who viewed the build-up of Soviet naval forces as the logical extension of the USSR's primary strategic objective – to defend the homeland against attack during a war – which, while not desired, was a possibility. Since the Soviets' structure their entire armed forces toward war fighting, the role of the Navy in a war would be to prevent sea-based attacks upon the Soviet Union. According to MccGwire, in a European war the Soviet objective would be to take Western Europe, inflicting as little destruction as possible to the industrial base, thus making the conquest worth the cost. Western sea-based nuclear forces, however, threatened to raise the cost by striking from secure positions at sea at the Soviet homeland. Thus, according to MccGwire, the Soviets have been compelled, first by the West's carrier attack capability and later by the introduction of SSBN, to push out their maritime defense zone. Generally, this entailed the ability to sustain naval operations in the Norwegian sea, northwest Pacific and eastern Mediterranean. The defensive rationale would also explain the more forward Soviet naval exercises. In order to meet the SSBN threat, the Soviet Navy

> would need to develop the capability to sustain deployments in the vicinity of US naval bases, to cover the main routes from North America to the Mediterranean and Norwegian Sea, establish infrastructure in other areas of threat which would allow for the rapid establishment of counter measures should need arise.
> (MccGwire 1975: 513)

This view cast serious doubts on whether the Soviet Navy could meet its defensive objectives in more distant waters. Whatever its capability, its expansion was nevertheless 'roughly in proportion to the growing degree to which the United States and NATO have increased their own reliance on seaborne weapons system' (Erickson 1971: 60). The Soviet response to this Western threat has placed the waters surrounding Europe in a situation of no assured command for either side. 'This last development,' notes MccGwire, 'has brought particular discomfort to the West with its vital interest in the use of the sea. And it is relevant to ask whether the West has in fact, gained more from their past initiatives than they have lost from the Soviet response.'[7]

With the development of Soviet SSBNs, the importance of securing a wider maritime defensive zone increased, particularly in the Norwegian Sea. In the event of war, the SSBN fleet would remain in the Barents Sea to be used, if necessary, for intercontinental strike, continental strike or to be held as a national strategic reserve. The protection of this force meant that enemy naval forces could not be allowed to take command of the Norwegian Sea. In this sense, the Soviet defensive objective could require offensive action against Western naval forces. It might also entail a ground assault on northern Norway itself, in order to forward base aircraft and help secure access for naval forces which could contest control of the Norwegian Sea (MccGwire 1980: 181).

The offensive, global interpretation of Soviet naval developments does not dispute the original defensive intent of the growth of Soviet seapower. It argues, however, drawing heavily upon the writings of Admiral Gorshkov (1979), that the Soviet Union had, by the early 1970s, changed 'from being more or less a protector of Soviet coastlands' to 'a deep water Navy with global reach and objectives'. The new Soviet Navy was meant to be a 'political and diplomatic instrument the effectiveness of which cannot be gauged by its intrinsic ability to stand up to US naval power, especially in an extended conflict' (Kilmarx 1975: 5, 61). In his 1975 study, *Military Power and Soviet Policy*, Thomas Wolfe noted that the Soviet Navy was 'in transition from a force optimized to exact a high toll from US nuclear naval units at the outset of a general war, and thus to limit damage to the Soviet homeland, to one possessing a better world-wide general purpose naval capability' (Wolfe 1975: 51).

One variant of the offensive interpretation regards the development of the Soviet Navy into a 'blue water' navy not as the result of a grand design on the part of the Kremlin, but rather the result of bureaucratic politics and institutional perspectives. Senior naval officers have longed for a 'dream navy', a global fleet similar to the United States Navy, which can go anywhere and do anything. They have sought such a navy, 'whether or not such a navy is specifically desired by the Soviet leadership or really necessary for defensive or geo-strategic reasons'. According to this interpretation, MccGwire is wrong in attributing Soviet naval developments largely to reaction to Western developments. Western naval developments alone cannot explain the Soviet naval build-up. The Soviet naval leadership greatly expanded the USSR's naval capabilities according to its own 'professional self image' (McGruther 1978: 3–6).

Whether the development of the Soviet Navy reflected the desire on the part of the USSR's leadership to project political influence far from the Soviet shores, or whether much of that development merely represented the desire of Soviet naval leaders for a global fleet, by the early 1970s the Soviet Union was indeed employing its naval forces throughout the Third World, and maintaining a presence on the world's oceans. Nevertheless, it was the defensive aspects of Soviet naval development, as stressed by MccGwire and others, which posed the greatest threat to NATO's maritime posture. The implications of the Soviet Union's naval build-up for the Atlantic Alliance must be viewed in the context of a war or crisis. In a deteriorating East–West environment how would the Soviets dispose of their naval forces? Even the proponents of the offensive interpretation seem to agree that in a war with the West, the Soviets would concentrate their forces in their immediate coastal waters, particularly in order to defend the SSBN fleet.

This is not to suggest that in a NATO and Warsaw Pact war, of whatever magnitude, the Soviet Union would not seek to employ naval forces outside the area covered by the North Atlantic Treaty, or beyond the northwest Pacific. In the Indian Ocean, for example, several Soviet SSNs and SSGNs could present a serious threat to Western oil supplies coming out of the Persian Gulf. But, the major threat to NATO will result from the pushing out of the Soviet Union's maritime defense zone into the Norwegian Sea, North Sea and possibly the North Atlantic, out from the eastern Mediterranean, and certainly in the Baltic. Put simply, the pushing out of their maritime defense zone entailed contesting control of these waters with the NATO navies. Unless NATO controlled the ocean approaches to Europe, maritime support for allied forces on the ground through reinforcement and resupply and force projection could not be assured.

The extension of the maritime defense zone could well, even as proponents of the defensive rationale for Soviet naval development point out, entail more than the protection of vital Soviet bases against attack by the West. Having secured control of the seaward approaches to these bases, the Soviet Navy could then move out to contest wider ocean areas vital to NATO. All this was possible in light of the growth of the Soviet Navy, in particular the northern fleet facing NATO's northern flank. John Erickson noted: 'Even allowing for defense in depth of vital Soviet bases, the overall capability (in the north) outstrips a purely defensive requirement and goes far beyond what is understood by the term "a flank"' (Erickson 1976: 67).

From the Soviet viewpoint, the ability to contest the West for control of the seas well beyond the immediate approaches to the Soviet Union constituted simply an extension of the fundamentally defensive role of the Soviet Navy. To NATO, this ability to push the maritime defensive zone further and further out, entailing, as it might, a ground invasion of northern Norway, necessarily appeared as a dangerously new offensive factor in the calculation of the balance of power, not only because NATO had to control these waters, but because the allied perception was itself one of defense. While the presence of the allied, and especially American, navies in the Norwegian Sea and the eastern Mediterranean would pose a direct threat to the Soviet homeland, and the Soviet Union's SSBN strategic reserve, the movement of large numbers of NATO naval forces eastward was anticipated as part of the NATO naval counter-offensive in reaction to some Soviet act of aggression or threat of aggression. NATO naval leaders did speak of the need to strike at Soviet bases. Nevertheless, the basic assumption underlying the use of NATO naval forces, and indeed of all allied forces, even those which, unlike naval forces, were in peacetime under allied command, was that these forces would come to the aid of those member states or state which was attacked or threatened with attack. Moreover, the use of these forces in a NATO capacity could only come subsequent to a collective decision by the allied governments as expressed through the higher political authority. Here again, the assumption of reactive and defensive use was emphasized. Even the deployment and actions of a combined fleet would be governed by the higher political authority to the extent circumstance allowed. This would especially be the case during a crisis period wherein the several NATO governments would be seeking some kind of negotiated settlement to avoid war, even as the Alliance mobilized.

Thus, the implications of the Soviet Union's naval development for NATO can only be properly evaluated with reference to the possible missions undertaken by the Western navies and the varying circumstances under which seapower would come into play in the context of growing tension or war. An extended Soviet maritime defense zone would entail a contest for control of the seas. But the broader strategic importance of that contest, and the meaning of its outcome for either side, would turn on a number of factors, especially the escalatory level of the land and air war in Europe.

At the most extreme level of escalation, strategic nuclear war, the growth of Soviet naval capabilities, with the exception of the SSBN fleet, would have little impact. NATO would be able to hold its own SSBNs far out to sea where Soviet ASW would be extremely difficult. If both sides withheld their SLBMs, the Soviets would, as noted, want to protect their force in the Barents Sea from attack by NATO forces moving up the Norwegian Sea, especially attack carriers. Yet, if NATO or the United States or Britain, acting unilaterally, wanted to strike at the Soviet SSBN fleet in the context of a war that had quickly escalated to the strategic nuclear level, it would be done with long-range missiles without prior establishment of absolute control along the northern flank.

The Soviet concern for the protection of their sea-based deterrent would be more of a threat to allied control of the seas in the context of a crisis or war which escalated more gradually. The movement of NATO task forces into the Norwegian Sea and Eastern Mediterranean, either as a show of strength during a crisis or as a means to support allied forces ashore (with amphibious landings and carrier-based aircraft), would not likely go unchallenged by the Soviets for very long. Even if the allies were only intending to support the defense of NATO territory and sea lines of communication, the Soviet Union would move to protect its homeland bases and might strike at the task forces rather than take the risk of having significant forces moving too close. Nor would the Soviets want to have the option of moving their own fleet out further foreclosed by the presence of allied navies.

It would be at this point that the enhanced Soviet anti-ship capability would threaten NATO naval forces in the fulfillment of their tasks, as non-escalatory as those tasks might be, i.e. reinforcement of northern Norway or securing the Baltic approaches. The contest for control of the sea might well touch off an 'inadvertent nuclear war', as the Soviets, fearing for their position, employed nuclear weapons against the NATO ships, the carriers in particular (see Posen 1982). According to one study of the Soviet Navy's 'declaratory doctrine for theatre nuclear war', the Soviet Navy 'makes little differentiation between the conduct of theatre naval warfare at the conventional or nuclear level'. Moreover, the 'pattern of Soviet naval operations does not necessitate a marked shift from a conventional to a nuclear mode and has inherent in it a considerable degree of flexibility' (BDM 1977: 42).

Another consideration is the nearness of the crucial sea areas to the Soviet Union. While it is true that in order to use their surface and subsurface naval forces against NATO the Soviet Navy would have to pass through choke points where their movement could not go unnoticed by NATO, the main striking power of the combined allied fleets would, in the case of the northern flank, first have to get up there before it could challenge the Soviets. Thus the problem of warning time, mobilization and deployment became a salient factor in assessing the magnitude of the Soviet naval threat to NATO. Even if the American underwater SOSUS system and satellite surveillance detected large Soviet naval movements, it would take several days to mount an allied response either under Major NATO Commands (MNCs, e.g. SACLANT), or, in advance of a decision by the North Atlantic Council, through the deployment of national forces under national commanders. One analysis concluded that: 'Unless US Second Fleet units are operating in the waters east of the Icelandic coast at the time of the attack, the Soviet Navy could, with relative ease, gain control of the Norwegian, Barents and Greenland Seas' (Davis and Pfaltzgraff 1978: 38).

In the Baltic, the Soviets could bring superior quantitative naval forces to bear without passing through a choke point. Only in the Mediterranean, where the US Sixth Fleet was supported by the Italian Navy, did NATO have a large peacetime presence available for war on short notice. Throughout the NATO area, allied surface ships would be vulnerable to Soviet land-based naval aviation operating from fields within the Soviet Union or Eastern Europe.

The pushing out of the Soviet maritime defense zone would then, in the event of war, particularly a conflict that did not quickly escalate to the strategic nuclear level, involve a battle for control of waters surrounding NATO Europe. However, the growth in Soviet naval capabilities suggested the possibility that the Soviet Union might extend its naval operations to include a major campaign against trans-Atlantic shipping, especially American reinforcement and resupply. Would the allied powers be faced with a third Battle of the Atlantic, or, as Admiral of the Fleet, Lord Hill-Norton, puts it, 'a battle for the Atlantic bridge' (Wilmott 1981: 151)?

On Soviet intentions and capabilities to conduct a campaign against NATO shipping, there is considerable disagreement in the academic and professional literature. The major Soviet naval exercises of the early 1970s did see the Soviet Navy engage in simulated anti-SLOC measures in the central Atlantic. Yet some analysts have suggested that such exercises represented a kind of psychological warfare: making the West think twice about taking steps in peacetime which would be threatening to Soviet interests by suggesting Western vulnerability (Daniel 1978: 227–8). Others have noted that, because the primary role of the Soviet Navy would be defensive, any attempt to cut NATO sea lanes would be only to draw allied naval forces away from waters nearer to the Soviet Union. Thus, MccGwire argues that although allied merchant ships 'are most certain to be attacked at the outbreak of war', this would only be a tactic on the part of the Soviets to pin down Western forces and 'more important, of diverting them from assaulting the northern fleet's inner defense zone and attacking the ballistic submarines therein' (MccGwire 1980: 170). In his *Guide to the Soviet Navy*, Siefried Breyer contends that the 'aim of the Red Fleet is to reduce the offensive power of the Western allies by forcing them to commit far greater forces to the defense of the sea lanes than the Soviet Union commits to their attack' (Breyer 1970: 107).

Most professional allied naval officers, particularly those who served with SACLANT, argued that with their increased overall capabilities, the Soviets were able to think more in terms of anti-SLOC operations as a priority in addition to defense. Moreover, they noted that, from a NATO perspective, Soviet interdiction potential was most important 'because that is the dominant warfare capability built into the Soviet naval forces' (Swartztrauber 1979: 114).

With over 200 torpedo attack submarines and over 60 cruise missile submarines, significant numbers of subsurface vessels could well be dedicated to anti-shipping operations while the rest remained to defend the homeland and SSBN fleet (McGruther 1978: 60–1). Less than 100 submarines, 'even if only conventionally powered, would certainly be capable of wreaking havoc on NATO SLOC'. Moreover, as McGruther argues, while the Soviet Navy might not want to undertake this mission, in a NATO war in which the bulk of the fighting is carried on by the Army, Soviet leaders may accede to Army demands that the Navy divert some forces to interdict the Western capability to reinforce the theater of operations (ibid.). In its survey of Soviet naval doctrine, the BDM study concluded that, although not the highest priority, the Soviet Navy would attempt to break out some forces beyond the GIUK gap, and the Baltic Sea, 'for offensive operations against those sea lines of communication and theatre ports and facilities which have a direct and immediate impact on the continental land campaign' (BDM 1977: 41).

In sum, as a 1980 Congressional Budget Office study observed, although the Soviet Navy is not 'optimized for an anti-sea lane strategy, its forces have significant capabilities to pursue such a strategy'. 'Furthermore,' the study continued, 'deployments appropriate to SSBN protection operation are also appropriate for sea lane attacks. It therefore appears prudent for Western forces to plan to defend the sea lanes against a Soviet Navy whose ability to attack those sea lanes is growing' (US CBO 1980: 38).

This capability would be greatest in the immediate European coastal waters. In these waters, the large numbers of smaller vessels, armed with surface-to-surface missiles, could exact a heavy toll on allied shipping as it crowded into northern European ports with military equipment destined for the front. In his study *The Red Navy at Sea: Soviet Naval Operations* Bruce Watson argues that the development of the smaller vessels, such as the Nanuchka Class missile patrol boats, 'suggests that the Soviets may intend to use these fast, heavily armed, lightweight craft in coastal areas, thus permitting the use of guided missile equipped destroyers and larger ships on the open seas' (Watson 1982: 32).

The flow of allied shipping could also be threatened by Warsaw Pact mining in the vicinity of major debarkation facilities. In addition to their fleet of minelayers, the Soviets and their allies were expected to employ naval aircraft, fishing vessels and merchant ships to block Western European harbors. Allied naval officers also note that, in the context of a still limited conflict, in which there was some naval action but no major attacks on shipping, the Soviets might sabotage the entrance to ports by deliberately sinking vessels in narrow channels.[8]

Attacks on allied reinforcement and civilian shipping operations in the coastal areas could be supplemented with amphibious landings. Such landings could be undertaken to secure shore areas as an additional means of striking at the SLOC, i.e. the Danish narrows, or to support the movement of Soviet land armies. The amphibious capabilities of the Warsaw Pact did not by any means match those of the US. But, here again, the Soviets would have certain advantages such as shorter distance to travel in order to land forces on NATO shores. As one study commented: 'Given the importance of these regions to the defense of NATO ... it would be imprudent to discount completely Soviet amphibious capabilities' (Blechman 1973: 15).

Evaluating the Soviet Navy by the mid-1970s from the perspective of NATO maritime interest it cannot be said with absolute assurance that the Soviet Union had the capability to win the war at sea. Yet it seems evident upon reviewing the professional and academic literature that the number of Soviet ships and aircraft, and their fighting capabilities, represented 'a significant shift in the balance of maritime power by ending the era in which the United States and its European allies could confidently use the seas without fear of effective challenge by the Soviet Navy' (Till *et al.* 1982: 68). In terms of being able to control the seas and exploit their strategic value for its own purposes, NATO had moved from the assurance of working control of the seas, to a situation where the Alliance had to anticipate disputed control in the event of war. Both sides would now operate at considerable risk. In such an environment, the NATO navies would be compelled to settle for working control over limited sea areas and for limited times in order to conduct specific operations, rather than waging an all-out campaign to achieve maritime mastery.

For the Soviets, this situation appeared to open up 'a new range of options in both war and peace', options which were 'particularly important because of the strategic and economic dependence upon the free use of the seas of its main rival, NATO' (ibid.). They were important not only because of the Alliance's continued dependence on the sea, but also because of NATO's new strategic doctrine, flexible response.

FLEXIBLE RESPONSE: THE MARITIME IMPLICATIONS

There seems little doubt that expectations of a protracted conventional or limited nuclear land war in Europe would serve to heighten NATO concern with the growth in the USSR's maritime capabilities. Allied naval leaders were foremost amongst those issuing warnings to the Alliance's political leaders. There was, to be sure, a measure of self-interest on the part of the allied navies, the US Navy in particular, in stressing the possibility of a protracted conflict with the Warsaw Pact. As Dov Zakheim, of the US Congressional Budget Office, noted in a 1976 study on naval force alternatives: 'A Navy strategy which presumes a prolonged conventional war in the Atlantic/Mediterranean region generates requirements for large forces.... This assumption could represent unrealistic expectations of Soviet intentions or NATO capabilities.'[9]

As much as it would have served the narrow service interest of the USN and other allied navies to hold to these assumptions, and whether or not they were based on unrealistic expectations, the Alliance's political leadership adopted in December 1967 a strategy which suggested the possibility of a protracted conflict – a conflict in which seapower would play and important role. At its meeting that month, the North Atlantic Council approved a 'revised strategic concept' submitted to it by the Military Committee, as set forth in MC 14/3, and 'following the first comprehensive review of NATO's strategy since 1956':

> This concept, which adapts NATO's strategy to current political, military, and technological developments, is based upon a flexible and balanced range of appropriate responses, conventional and nuclear, to all levels of aggression or threats of aggression. These responses, subject to appropriate political control, are designed, first to deter aggression and thus preserve the peace; but should aggression unhappily occur, to maintain the security and integrity of the North Atlantic Treaty area within the concept of forward defense.
> (NATO 1982: 197)

The difficulties with flexible response are well known and will not be discussed at this point. Certainly, NATO's political and military leaders are well aware of its shortcomings. Yet, since there could be no turning back to massive retaliation, it was necessary to seek ways and means of improving the Alliance's sub-strategic nuclear posture.

The strategy of flexible response can be summarized from Schlesinger 1977 as follows:

1 Any aggression short of a general nuclear war will initially be met with a direct forward defense at the level, nuclear or conventional, chosen by the aggressor.

2 In the event of a conventional attack, NATO forces will attempt to wage a conventional defense, to raise the nuclear threshold. NATO will, however, resort to the first use of nuclear weapons if conventional defense is insufficient to halt a Soviet conventional invasion which threatens to overrun large amounts of NATO territory.
3 In its use of nuclear weapons, NATO will emphasize military targets and set as the primary objective for the use of nuclear weapons the termination of war on terms acceptable to the United States and its allies at the lowest feasible level of conflict.
4 NATO will initiate an appropriate general nuclear response to any major nuclear attack.

This approach to deterrence and defense is based upon certain explicit and implicit assumptions:

1 While all NATO territory, and all waters crucial to NATO, fall within the protection of its combined armed forces, the central front is considered the most important. It is where the Soviet Union and its allies have concentrated their forces. Flexible response is particularly important along the central front because of its escalatory potential, yet, it is here where flexible response will be especially difficult because of the concentration of forces. Limited Soviet incursions would be more likely along the NATO flanks, particularly at the Soviet and Norwegian border. Here a flexible, conventional resistance would also carry a comparatively lower risk of escalation to the nuclear level.
2 There will be no strategic surprise on the part of the Warsaw Pact. Conflict, at any point along the Alliance's frontier, or at sea, and of whatever magnitude will come about following a period of rising tension accompanied by Warsaw Pact mobilization.
3 In order that the Warsaw Pact does not achieve 'tactical surprise' NATO must use its warning time to mobilize as well, keeping pace with the Pact mobilization (Karber 1970: 286).

NATO must mobilize and reinforce before the outbreak of hostilities and be able to sustain reinforcement and resupply after armed conflict begins. Although estimates vary as to how long the standing forces could hold out against a major Soviet conventional assault on the central front, it is generally accepted that: 'The conventional defence of Central Front depends crucially on transatlantic reinforcement and resupply' (UK Ministry of Defence 1981: 25).

On the NATO flanks, it is generally assumed that the standing forces would be able to offer little resistance and therefore must be reinforced at the earliest possible moment after a NATO decision to mobilize.

The strategy of flexible response and the assumptions which underlay it had a direct and significant impact on the role of seapower in the allied deterrent and defense posture. Flexible response heightened NATO's need to exploit the strategic value of the seas, and therefore the necessity of controlling the seas.

In the first and most obvious instance, the Alliance had to control the seas in order to bring in those reinforcements and supplies upon which conventional deterrence and defense rested. Despite prepositioning and airlift, 'the bulk of equipment and resupply would have to come by sea' (ibid.). As the US Defense Department noted, sea control would assume 'its greatest importance in an extended, conventional European war'. Reinforcement and resupply would have to begin prior to the outbreak of hostilities and extend for an indefinite period after the war had begun (US CBO 1976: 4).

If a NATO and Warsaw Pact conflict should quickly escalate to the nuclear level, 'even if nuclear weapons are confined to the immediate theater', sea control for reinforcement and resupply would become less crucial. As a United States Congressional Budget Office study noted: 'the destruction of men and material, and consequent shortage of supply, may well be so great as to be beyond the capacity of convoys to replenish' (ibid.). Moreover, the convoys themselves, as well as port facilities, would be likely targets of nuclear attack.

This is not to say that in the event of unilateral or mutual escalation to nuclear weapons, NATO would abandon all efforts at reinforcement and resupply. The scope of nuclear usage and political decisions bearing upon the subsequent response of NATO would dictate the level and direction of allied shipping. It has been suggested, for example, that should the major harbors in northern Europe become unusable, shipping would be directed toward British ports on the Atlantic.[10] Yet, clearly, flexible response, with its implied use of nuclear weapons against a large-scale Soviet conventional assault which threatened to overrun Europe, recognized the limits of reinforcement and resupply at sea.

At the same time, this strategy, with its emphasis on conventional response, and the ambiguity of its first use doctrine, was placing a great deal of reliance upon reinforcement and resupply. To quote again from the CBO study mentioned above:

Should a conventional war extend past a month, for whatever reason, and however great expectations of its duration may be to the contrary, its successful conclusion could then critically depend upon the [US] Navy's ability to ensure safe supply by controlling vital sea lanes.

(US CBO 1976: 4)

The control of vital sea lanes for the purposes of reinforcement and resupply would also afford NATO a better chance of avoiding having to make a decision regarding the first use of nuclear weapons in the event that the Soviets did not use them initially. Should NATO find its sea lanes cut, the allies could be faced with two unpalatable alternatives: a back-to-the-wall nuclear response, or capitulation, thus imposing the requirement in peacetime of sustaining the Alliance's conventional naval capability as means of further strengthening deterrence. A Soviet understanding that if they launched a conventional attack NATO could bring in reinforcements would serve to dissuade the Soviet Union from such an attack (Swartztrauber 1979: 115).

NATO's ability to control the seas would also allow for the use of force projection ashore as an additional means of implementing a flexible response. As noted in previous chapters, carrier-based airpower had been considered as another source of tactical air support for SACEUR. This would be particularly true on the southern and northern flanks.

The implications of flexible response for the Sixth Fleet in the Mediterranean were quickly recognized by Admiral Horacio Rivero, USN, who became CINCSOUTH in February 1968. He changed the primary mission of the fleet, which became STRIKFORSOUTH upon the declaration of a NATO reinforced alert, from one of nuclear strike to support for the air and land battle in defense of Greek, Turkish and Italian territory. This task was to be subordinated only to the neutralization of those Soviet naval units which posed a direct threat to the carriers. According to Rivero:

> The relative force situation was such that success in the battle for territorial defense depended critically on establishing and maintaining local command of the air over the battlefield, something which the indigenous air forces could not do without a substantial infusion of US airpower in the earliest stages of the battle. The Sixth fleet carriers, plus deployed US Air Force squadrons at Avino, Italy and Adana, Turkey and the additional USAF squadrons that would immediately deploy from the US, were to provide this reinforcement of modern aircraft to supplement the less capable indigenous air forces. Since the immediate use of the deployed USAF squadrons could not be assured because of the QRA [Quick Reaction Alert] requirement for nuclear attack assigned to them (and from which the Sixth fleet squadrons were removed), I had to count primarily on the Sixth fleet to provide the

modern aircraft needed to influence the critical early stages of the land/air battle.[11][12]

Along the northern flank 'the requirement for naval power projection against attacking forces ... would certainly be significant' (US CBO 1976: 9). While the standing air and ground forces along the central front would not be greatly augmented by force projection, i.e. through amphibious landings or through carrier-based aircraft in Norway, which did not allow the stationing of foreign forces in peacetime, NATO anticipated a major reinforcement effort. This would be necessary in the event that, for whatever reason, the Soviets elected to launch a limited attack on Norway. In the late 1960s and early 1970s, a good deal of attention was given to force projection capabilities along the northern flank. At the United States Naval War College, the Commander Striking Force Atlantic conducted the ESTABLISH CONTACT war gaming exercises in conjunction with representatives from the northern European commands of NATO. The exercise involved a 6 carrier striking fleet (4 American, 2 British) trying to get to Norway 'opposed by the Soviet threat'. Its purpose was to increase the ability of various NATO commands to function together and demonstrate 'what the Striking Fleet can do for Norway'.[13] In the 1972 'real life' exercise STRONG EXPRESS, a British and an American carrier task force provided air cover for the landing of a combined amphibious force in response to a limited Soviet conventional attack on northern Norway.[14]

In addition to expanding the role of seapower in conventional warfare, flexible response also raised the possibility of a wider use of naval forces in the event that NATO elected to engage in tactical/theater warfare. The US Navy's carriers retained a nuclear strike capability. Both Britain and the United States had earmarked a portion of their SSBN fleets for SACEUR's possible use. The introduction of the MIRVed Poseidon SLBM, with its lower-yield individual warhead capable of more discriminating strikes, afforded SACEUR supplemental theater nuclear capability. Nuclear weapons on board American surface ships also afforded NATO the option of extending its nuclear response to the war at sea, instead of, or in addition to, use of nuclear weapons in the air and ground war. In the course of ESTABLISH CONTACT, procedures were tested for the 'selective and pre-conditioned release of authority for nuclear weapons'.[15]

Not only did flexible response mandate an expanded role for maritime forces once hostilities began, and throughout the conflict spectrum up to strategic nuclear war, but the NATO strategy also placed a premium on the early pre-hostilities mobilization and deployment of the Alliance's maritime forces. Without this early movement NATO would not be in a favorable position to secure and exploit the seas. Along the northern flank the pre-hostilities movement of maritime forces had to complement the mobilization and deployment of NATO ground and air forces earmarked for Norway. Indeed, unless the Alliance was able to secure Norwegian territory, the Soviets would be in a more favorable position to contest control of the seas. By securing important forward positions in Norway, the Soviet Union could deploy more effective land-based air cover for its naval forces.

In exercise STRONG EXPRESS, NATO implemented a series of pre-hostilities deployments in response to a hypothetical strategic warning of a Soviet attack on Norway. The exercise assumed a two-week period between the first strategic warning and the actual attack, during which time the following steps were taken:

Sept. 3: Allied Command Europe (ACE) mobile force units in Italy, Canada, the UK, Germany (the USAF), Netherlands and Luxembourg are put on alert.

Sept. 11: Information is received of a definite attack and the first elements of NATO's Air Mobile Force begin deploying to Norway.

Sept. 13: NATO surface ships and submarines begin building up off the southern coast of Norway.

Sept. 15: The eastern Atlantic is covered by long-range patrol aircraft flown from the UK, Iceland, Netherlands and Germany. A British carrier task force, including amphibious forces moves toward Norway. An American carrier task force moves in from the Western Atlantic. Britain and Belgium begin antimine operations in the Channel area.

Sept. 16: Amphibious forces from the Netherlands, UK and US make unopposed landings. Minesweeping operations at the Baltic approaches. Continued airlift of ACE mobile force.

Sept. 18: The Soviets attack in Norway.[16]

In this exercise NATO sought to secure the seas to allow for projection of force ashore. Securing the seas during a time of strategic warning would also be necessary to ensure reinforcement and resupply subsequent to the outbreak of hostilities. Although NATO would want to begin sealift in advance of hostilities, reinforcements may not arrive in Europe in significant numbers by the early stages of a war; 'however, larger warning times allow for measures to be taken that would improve the prospects for early and efficient protection of the sea lanes' (US CBO 1977: 5). ASW forces would be moved forward, and bases such as on Iceland and in the Mediterranean could be augmented with additional tactical aviation to 'withstand initial attacks from Soviet aviation' (ibid.).

The mobilization and deployment of the allied navies and airpower dedicated to sea control would not only have to begin in advance of hostilities, it would have to proceed as rapidly as possible in order not to be caught by tactical surprise. This in turn meant that NATO's deterrent posture had to reflect a willingness and capability to attend quickly to the maritime aspects of flexible response:

> A flexible response strategy meant nothing if it did not mean the demonstration in peacetime of both the determination and the capability to engage major enemy forces by land, sea and air if put to the test. Given the speed of movement and the scale of attack which had to be expected, as well as the range of missiles, the whole land and sea area of the north-west European continental shelf would have to be regarded, from the outset, as part of the central front battlefield.
> (Hackett *et al*. 1978: 344)

Indeed, given a rapidly moving battle on the ground and the consequent need for early reinforcements of a magnitude only possible through sealift, a flexible response could require that NATO would have to attempt to reinforce without awaiting the outcome of the war at sea. As Geoffrey Till noted, 'although it may be safer for NATO to remove Soviet naval threats before the reinforcement convoys sail, time and events on the European mainland may not allow stately procedures' (Till *et al*. 1982: 191). Not only the course of the land battle, but the very nature of the reinforcement and resupply effort would require convoying prior to the determination of the war at sea. As noted above, the sealift would have to have something in it on D-day. Thus, some military convoys, and perhaps crucial civilian economic shipping such as oil tankers, would be at sea when the fighting started.

A final, but not often mentioned, maritime implication of flexible response was the increased possibility of a NATO and Warsaw Pact confrontation either initially being confined to the sea or escalating faster at sea than in the land and air war. In Chapter 3, reference was made to American war-at-sea planning. This involved the development of contingency plans to fight limited or major wars, conventional and nuclear, with the Soviets at sea with no concurrent large-scale land war. It was believed that, given its continued edge at sea, the risks of escalation on land and the danger to allied civilian populations inherent in any land war, the

United States might elect to respond to a Soviet provocation by first striking at Soviet ships at sea, both naval and merchant. At its height, USCINCLANTFLT's war-at-sea planning postulated a full-scale confrontation between the American and Soviet fleets in the North Atlantic.[17]

The arguments advanced for war-at-sea preparations fit very well into the assumptions and purposes of flexible response. As Wilson and Brown suggested in a 1971 study for the US Navy's Center for Naval Analyses, the West might well find it to its advantage to have an initial military confrontation with the Soviets begin at sea. At sea, the war would be 'inherently more easily limited ... militarily more decisive on a campaign basis' and easier for combatants to control their actions during hostilities and to terminate hostilities (Wilson and Brown 1971: 5).

In a war fought exclusively at sea, for however long, allied maritime forces would be exploiting the strategic value of the seas primarily to locate and destroy Soviet naval units. Reinforcement and resupply and force projection ashore would not be the immediate purposes for the NATO maritime forces (although, having won the sea battle, the oceans could be exploited to achieve these objectives). Nevertheless, even in a war confined exclusively to the seas the ultimate objective would lie ashore, namely the territorial integrity and security of the European allies. Ideally, allied security would be better served if, in implementing a flexible response, NATO could confine itself to the ocean areas and inflict sufficient losses upon the Soviet Union so as to prevent a land war. Whether the Soviets would accept a defeat at sea without launching a subsequent land attack, or whether they would not engage in a concurrent land assault, especially against Norway where conquest of territory would improve their position at sea, was doubtful. The notion of initial hostilities, perhaps even nuclear, being confined to the seas was, however, not beyond the realm of possibilities.

In summary, it can be said that the adoption by NATO of flexible response increased the Alliance's need to exploit the strategic value of the seas. It did so spatially, temporally and operationally.

Since NATO wanted to deter and defend against a wide spectrum of possible Soviet actions, including limited conventional incursions along the flanks, much more sea space became crucial. Under a massive retaliation posture, the Alliance would not necessarily have to allocate many resources to defend the immediate coastal waters. Nor was the naval balance in the Baltic particularly important. A strategic nuclear attack could be launched from the US mainland or from British or French soil. Seapower would be important if SSBNs and nuclear armed carriers participated in the strike, but in these instances NATO would need only limited 'working control' of the seas around the weapons platforms to allow for launching. To be sure, in a pre-hostilities period it would be important to secure sea-based nuclear weapons platforms from pre-emptive attack, but there would be less of a need for defense of sea lanes of communication, or for the coastal waters that gave access to the land areas, where a conventional war would be fought. Under flexible response, these logistic avenues became crucial (Morse 1977: 22).

This concern with the SLOC extended the sea areas of importance further back across the Atlantic and throughout the world. While it was generally agreed that the main Soviet naval threat would come close to the Eurasian land mass, Soviet naval movements beyond the immediate waters could not be ignored. Cuban and African bases afforded the Soviet Union a wider surveillance capability as well as an enhanced capability to strike on the wider oceans. Western concern with oil shipments could not be ignored.

Temporally flexible response demanded a pre-hostilities mobilization of allied maritime forces in order to commence surveillance and to be in a position to meet the Soviet attack if and when hostilities began. The securing of sea control had to continue for as long as necessary and was dependent upon the nature of the land war and political decisions regarding allied responses. The greatest demands upon the Alliance's collective maritime forces would come in the event of a conventional conflict which lasted longer than a few weeks or in a limited nuclear confrontation. Since NATO could not know for certain what the length of a conflict with the Soviet Union would be, flexible response necessitated greater assurance that the Alliance would be able to exploit the strategic value of the seas from the outset and for an indefinite period. As the then CINCSOUTH, Admiral Horacio Rivero, USN, noted in 1972, in reference to flexible response and the allied posture in the Mediterranean: flexible response meant 'readiness on D-day' and 'staying power' (Rivero 1972: 7).

In terms of the scope of operations, 'readiness on D-day' and 'staying power' flexible response was consistent with the efforts of allied naval leaders since 1949 to maintain an as broad as possible capability. They had long held that the uncertainties associated with a future war involving the Soviet Union in Europe (uncertainties as to the use of nuclear weapons, as to the Soviet intentions and as to the responses of the Alliance) required that the allied navies maintain the capability to meet as wide a scope of scenarios as resources would allow (Till *et al.* 1982: 184–5).

Not only did flexible response explicitly call for meeting an attack at any level, the uncertainty as to what exactly the NATO response would be meant a broadening in the range of possible maritime contributions to deterrence and defense. Thus in the late 1960s, the attack carrier again assumed a conventional war support role while maintaining a tactical nuclear capability. Submarine-launched missiles were brought into the theater nuclear equation. By the early 1970s, NATO began to revise its plans for the mobilization of allied merchant shipping and convoying. A re-emphasis on shipping also raised the problem of harbor defense and countermine measures. Finally, flexible response suggested the possibility of a major NATO and Warsaw Pact conflict confined to the seas.

Whether conducted in the absence of concurrent land campaigns or, as would be more likely, in conjunction with conflicts along the European fronts, the war at sea would witness maritime forces in combat for control of the sea. The battles at sea would be multidimensional confrontations in which combatants would rely heavily upon advance surveillance and communications (ibid.: 185).

It is precisely because flexible response increased NATO's need to exploit the strategic value of the seas, in terms of the overall strategic posture, that the continuing expansion of the Soviet Navy was so significant. For the Soviet Navy seemed to be broadening its maritime defense zone, taking in waters further and further out from the Eurasian land mass. Its capabilities, though not exceeding those of the allies in all areas, were growing steadily by the mid-1970s. Just as NATO began to look to its maritime forces to keep options open, so would the Soviets seek to use their seapower to close those options.

MARITIME COMMAND STRUCTURE CHANGES

Whatever the ultimate wartime implications of Soviet naval developments on the strategy of flexible response by the late 1960s, NATO's naval leaders had begun an effort to convince the North Atlantic Council and the individual allied governments of the need for greater emphasis upon the Alliance's maritime posture. In their presentations to political leaders, SACLANT senior staff members such as Richard G. Colbert, USN, pointed out the 'Soviet World Wide Maritime Challenge' (as one official briefing paper was entitled).[18] While naval leaders did speak in global terms regarding the growing capabilities of the Soviet fleets, their actual assessment of the Soviet naval threat was more modest and differentiated.

The Soviet naval threat was viewed as dual faceted, with a 'cold war' component and a 'hot war' component. The former involved utilizing the Soviet Navy to support broader Soviet objectives in the Third World in an attempt to outflank NATO. Since direct pressure on the European mainland was too risky in terms of escalation, Moscow had shifted to the Third World as a forum to continue the struggle with the West, and the Navy would play its role in this 'global political and economic strategy for progressive attainment of their goals of Communist World domination'.[19]

While this 'cold war' component of the Soviet naval threat did concern the individual NATO governments, and while the new bases afforded the Soviet Union to extend the reach of the Soviet Navy, it would be difficult to implement a NATO 'counter-strategy'. According to a letter written by Colbert to a Swedish officer in 1968, the Soviets could carry out their Third World activities 'avoiding a military confrontation with NATO, in such a way that little could be done directly to counter it'.[20]

The Soviet 'hot war' strategy, however, was different and something NATO could do something about in terms of the Alliance's peacetime posture. In a direct military confrontation with NATO, the Soviet Navy would carry out a combined defensive and offensive strategy – defensive meaning protection of the homeland and offensive entailing 'operations to cut NATO's sea lines of communication and strike with their seabased ballistic missiles against NATO land areas'. To carry out these tasks against the Alliance the Soviets were developing an increased capability. Depending on the course of the war, they could go in either direction or both. Given the large numbers of attack submarines, allied naval leaders believed that the Soviets would attempt to attack NATO SLOC while leaving sufficient forces to protect their SSBN by striking at Western fleets. In pursuit of their 'cold war' objectives, the Soviets would have more forces permanently at sea and this would make it difficult for NATO to interpret Soviet naval movements as a signal indicating preparation for general war. Yet, the key areas for NATO remained the waters around Europe.[21]

For allied naval leaders, therefore, the Soviet threat at sea arose mainly out of the challenge the new Soviet capabilities presented to the fulfillment of allied maritime tasks and other aspects of allied defense. To be sure, this was all that allied planners could concern themselves with under their operating authority. But it is evident that the likely deployments of the bulk of Soviet maritime forces in the event of a major East–West crisis would be in those waters of direct concern to the Alliance's naval leaders.

At the beginning of 1967, SACLANT's staff was pressing for a 'new concept of maritime strategy' for the Alliance. The basic thrust was to increase NATO's ability to respond quickly and collectively at sea in the event of a crisis and to affirm the importance of the sea for NATO's overall strategy.[22] Unlike the ground and air forces along the central front, forces assigned to the naval commands were not permanently under NATO command. While the principal naval commanders as national commanders had forces directly at their disposal, NATO had to be able to move collectively at sea in a short time. Thus, in 1967, SACLANT set out to persuade the North Atlantic Council to approve the creation of a standing force permanently under his command.

From 1965 to 1967, SACLANT had conducted the MATCHMAKER exercises which saw several allied ships carrying out an extended joint patrol with underway replenishment. This provided SACLANT's staff with a basis for evaluating the feasibility of 'forming a permanent NATO squadron'.[23] As further means of increasing the immediate availability of forces, the SACLANT staff also developed the idea of a 'Maritime Contingency Force' (MCF). This force would be drawn from forces already earmarked by national governments, but would be available on short notice to support SACLANT's contingency plans. Taken together, the permanent squadron and the MCF were viewed as 'among other things' providing 'a substantial part of the answer to the Northern Flank problem'. There, the growing Soviet presence might place NATO at a disadvantage in the early moments of a crisis.[24] In a November 1967 letter to a Finnish naval officer, Richard Colbert noted that: 'We have a new strategy that hopefully will be adopted [at the North Atlantic Council meeting] in December ... and from our SACLANT viewpoint we have made a lot of progress in selling NATO maritime strategy.'[25]

At the same meeting in which the flexible response strategy was endorsed, the North Atlantic Council did agree to transform the MATCHMAKER naval training squadron into a Standing Naval Force Atlantic (STANAVFORLANT). This force, to be composed of destroyer-type ships, would, by virtue of being continuously operational, 'enhance existing co-operation between naval forces of member countries' (NATO 1982: 197–8). Colbert viewed the decision as 'a great success for us in our efforts to strengthen the Alliance by reorganization and employment of existing naval forces already available in peacetime'.[26]

STANAVFORLANT was commissioned on 13 January 1968 at Portsmouth, England. The original composition included ships from the UK, Norway and the Netherlands. In subsequent years anywhere from 4 to 11 destroyer-type ships, and occasionally submarines, would sail with the force. An oiler also accompanied the Force. Member nations contribute ships as part of their normal training cycles, and command of the force rotates yearly. Formally under SACLANT's ultimate command, STANAVFORLANT would come under CINCEASTLANT's operational control when it was in European waters.

As described in formal briefings for US Navy officers,[27] STANAVFORLANT had four main functions. First, it would provide training experience for joint operations enabling allied navies to improve 'naval operational proficiency and NATO tactical development and evaluation'. Second, by its very existence, STANAVFORLANT would give evidence of allied solidarity. Third, the force would provide NATO with a multinational ocean surveillance capability to monitor Soviet naval exercises and movements.

The fourth function of STANAVFORLANT would be to provide SACLANT with an immediately available combined force to be deployed 'to the scene of any possible contingency to reaffirm the solidarity of the NATO alliance and provide a visible deterrent'. STANAVFORLANT would move 'quietly to a threatened area, or just out of sight over the horizon' and thus be ready to respond to higher political and military direction while at the same time exercising the right of freedom of the seas. The forces could also provide the initial elements around which a 'more powerful and versatile NATO force could be formed' if tensions continued to escalate. In sum, this multinational squadron was viewed as being useful as a low-level allied response, 'ready to be used in operations intended to deter Soviet aggression under a NATO strategy of controlled response'.[28]

Because STANAVFORLANT might be employed in a surveillance role, or as a very low-level collective response during the initial stages of a crisis prior to a general mobilization of allied forces, higher political and military control of forces was considered essential. The Military Committee and the NATO situation center in Brussels were to be kept fully informed of the 'schedule, movements and activities of the force'. To quote from a 1972 briefing on procedures to be followed in the event that SACLANT wanted to employ STANAVFORLANT in a surveillance capacity:

> When these opportunities arise to deploy the force, without serious interruption to STANAVFORLANT's scheduled operations, the Military Committee is warned by message of the developing situation as early as possible; included in the warning message is current intelligence on the Soviet maritime force movements and proposed rules of engagement to be in effect throughout the operation:
> A. Training of armament is forbidden
> B. Do not interfere with activities of units under surveillance
> C. Aircraft, including helicopter buzzing by flying directly toward or overhead at low level is forbidden

D. Use of helicopters is authorized
E. Approach designated units under surveillance no closer than one mile

Providing the intended surveillance mission materializes as anticipated, the warning message is followed by a message of intent advising the Military Committee of SACLANT's intention to conduct the operation. SACLANT is in turn notified by the Military Committee on whether or not there are military or political objections to the mission. When the Military Committee has agreed to the surveillance operation, SACLANT directs the MSC (major subordinate commander, i.e. CINCEASTLANT), in whose area the surveillance operation is taking place to conduct the operation for the period of time agreed upon by the Military Committee.

During surveillance operations the highest councils of NATO are kept fully informed and further, they participate in discussions pertaining to the employment of force. These evolutions have demonstrated that opportunities for special operations during peacetime are often fleeting. As a result, the Military Committee and Defense Planning Committee have formulated special decision making procedures in order to be able to react promptly to developing situations.[29]

Although restricted in the performance of close surveillance, STANAVFORLANT did serve a very useful function in training allied naval personnel in working together. In the years subsequent to 1968, with continual rotation of contributions, thousands of sailors passed through the force providing a cadre of officers and men familiar with allied procedures.[30] Moreover, as allied naval leaders continually argued, the real value of STANAVFORLANT was less its military strength, but its very collective nature. Such a multinational force contained a 'powerful political potential'.[31] It constituted a continuing affirmation of allied determination to meet the Soviet threat at sea and an early collective demonstration of solidarity in the event of a crisis.

There were limits, however, as to how effective a peacetime multilateral NATO fleet could be and the additional flexibility STANAVFORLANT actually afforded SACLANT in the event of a real crisis. The individual ships were still subject to their respective national governments' authority. As such, use of the entire fleet would still have to result from some kind of collective decision over which of the contributing nations would have an implicit special veto. The purposes of allied solidarity would not be well served if, in the event of a crisis, one or two contributing nations elected to withdraw their ships. The difficulties

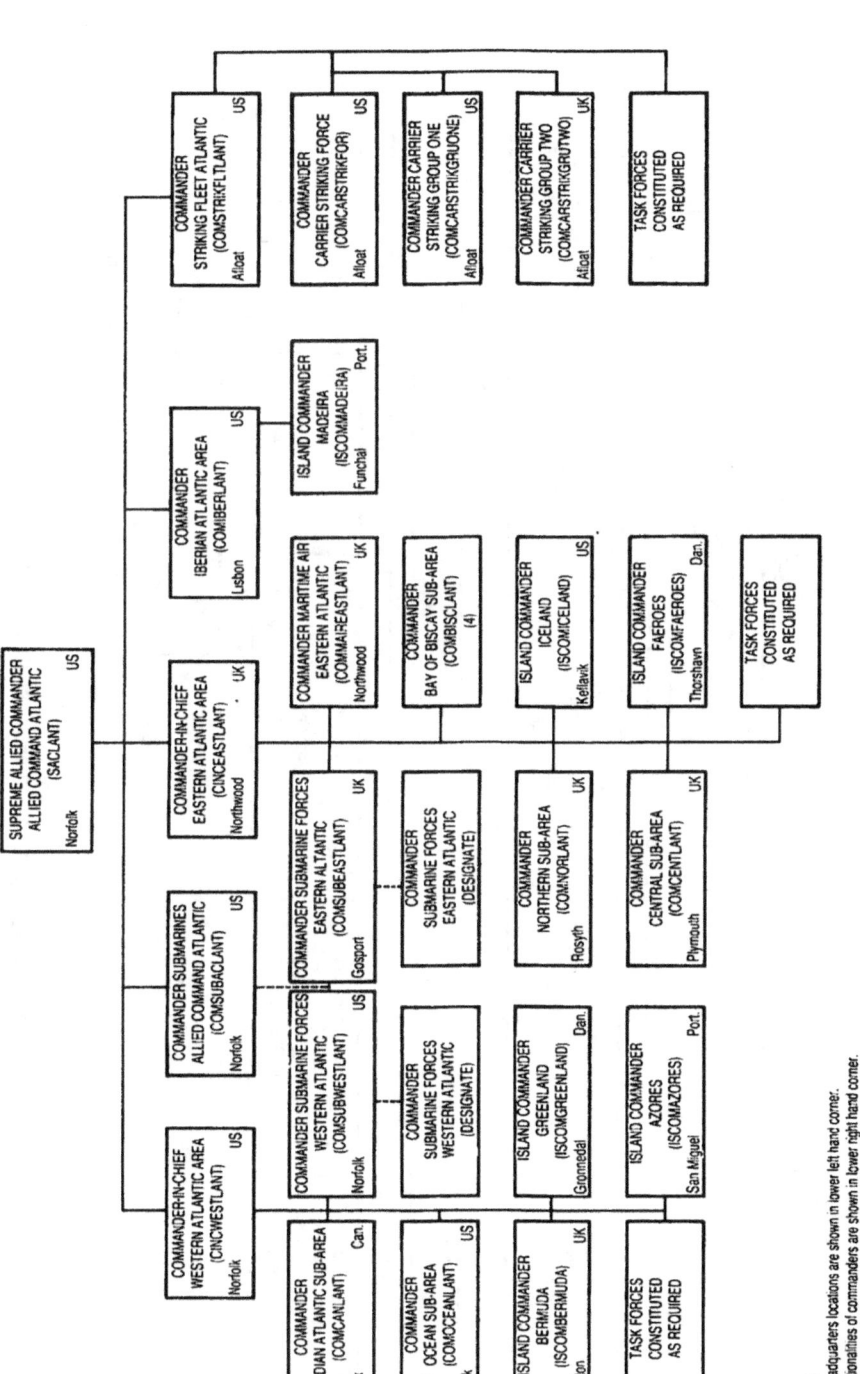

Figure 4.1 SACLANT in the 1970s

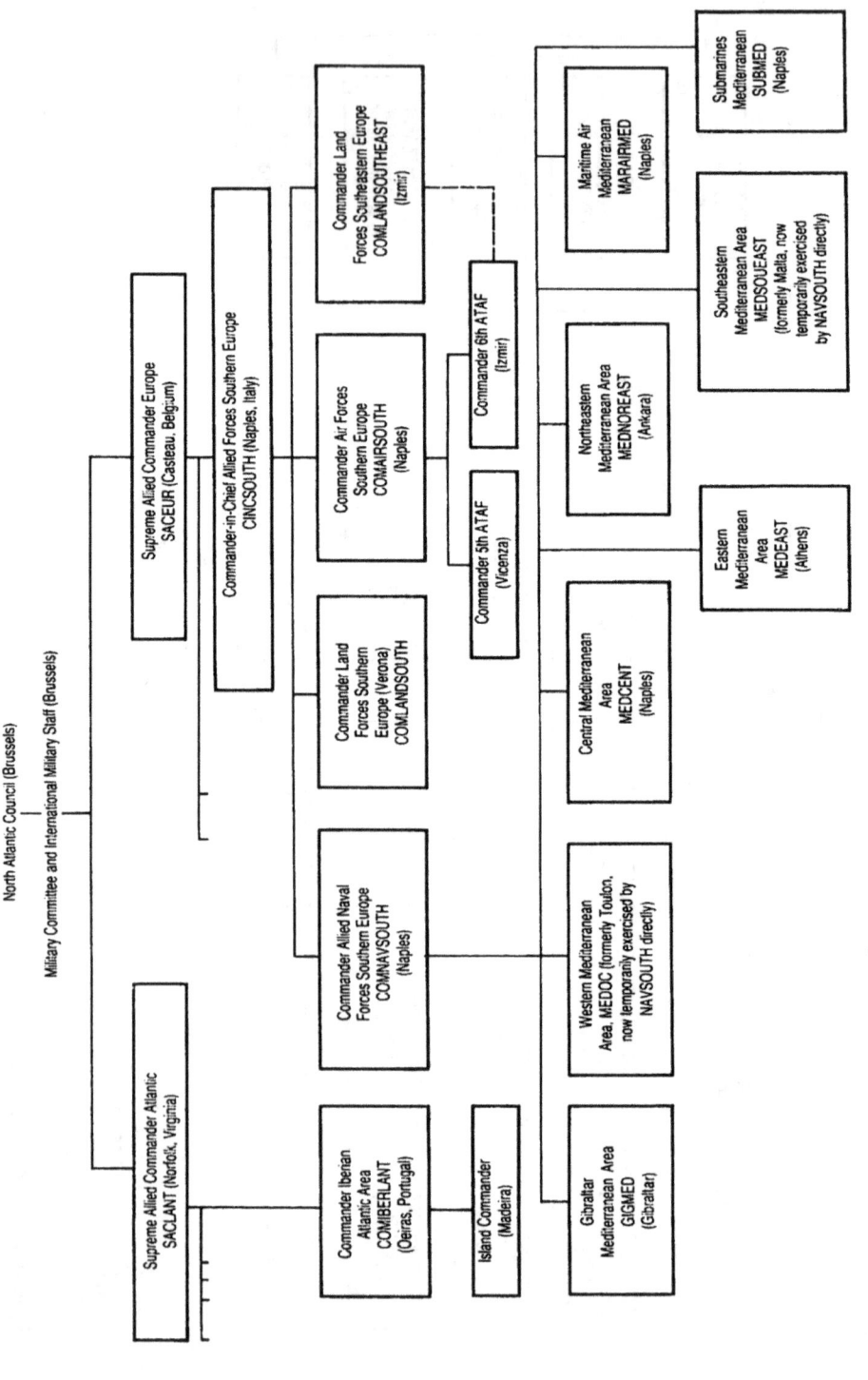

Figure 4.2 NATO military command structure elements in the Mediterranean and adjoining area

associated with the immediate effectiveness of the standing force were pointed out by US CNO Elmo Zumwalt when he rejected a suggestion that the United States nuclear powered merchant ship USNS *Savannah* be assigned to STANAVFORLANT as permanent, multinationally-manned command and control and replenishment ship. Included in the suggestion was a recommendation that the *Savannah* carry a mixed marine company composed of American and British marines available for quick reaction landings by helicopters.[32] Zumwalt raised doubts about 'another mixed manning experiment' (apparently in reference to allied objections to the multilateral force suggestions of the early 1960s). He also pointed out that there would be 'political problems' attendant with the use of a mixed marine force.[33]

The Maritime Contingency Force concept was more in line with the nature of NATO's maritime posture, which relied upon earmarked national ships being made available subsequent to a collective decision. As a supplement to the 'day-to-day availability' of STANAVFORLANT, the MCF consisted simply of 'specially tailored multinational task forces' which could be called up in advance of the bulk of earmarked forces. Pursuant to a decision by the Defense Planning Committee, the MCFs could be called up to support various contingency plans, or as part of a graduated mobilization as the Alliance moved from 'simple alert' to 'reinforced alert'. According to a 1972 statement by the then SACLANT, Admiral Charles Duncan, USN, the MCF was one of 'the primary tools that SACLANT has to carry out the NATO strategy of Flexible Response, short of general war'.[34]

In the Mediterranean, CINCSOUTH Horacio Rivero had, as noted above, changed the mission of STRIKFORSOUTH to one of support of the land battle. This was necessary because in the early stages of a conventional conflict the Greek, Turkish and Italian land-based air forces under AIRSOUTH would be unable to assure command of the air over the battlefield. However, with the increased presence of Soviet forces, the Alliance had to begin to think more about control of the seas, not only to allow for reinforcement and the protection of economic shipping, but in order to assure STRIKFORSOUTH of a secure environment.

NATO had taken an important step in this direction just prior to adoption of the flexible response strategy. On 5 June 1967 the position of Commander-in-Chief Allied Forces Mediterranean was abolished. All allied maritime forces (with the exception of the Sixth Fleet) were placed under a new command subordinate to CINCSOUTH. The mission of Commander Naval Forces South (COMNAVSOUTH) was to defend and control the sea lines of communication in the Mediterranean and to conduct submarine and ASW operations in the Black and Mediterranean Seas. In the initial stages of a war, COMNAVSOUTH's first priority would be to 'neutralize' the immediate naval threat to STRIKFORSOUTH.

The position of COMNAVSOUTH was held by an Italian Admiral who was first headquartered at the British facilities on Malta. Under COMNAVSOUTH, the sea area continued to be subdivided into subordinate commands under national commanders. The western Mediterranean (MEDOC), formerly assigned to France, was placed directly under COMNAVSOUTH. The continuing Greek–Turkish dispute (which came to a head with the Turkish invasion of Cyprus in 1974 and the subsequent withdrawal of Greek earmarked forces) disturbed maritime co-ordination along the southern flank. For example, the Turkish Navy agreed to operate under the general direction of CINCSOUTH, but would not agree to be placed under a Greek Admiral (COMEDEAST) when operating in the Aegean.[35] (Greek personnel continued to serve on allied staffs with Turkish representatives. Another problem resulting from the Greek–Turkish dispute was the arms embargo imposed on Turkey by the Congress with Turkey restricting port access by US ships.)

Although the British held the subordinate command at Gibraltar, by the early 1970s the standing British naval presence in the Mediterranean had been reduced to one destroyer, two frigates and two submarines with the occasional visit of a commando carrier. By 1976, the UK had withdrawn all naval units earmarked for NATO in the Mediterranean, concentrating its maritime contributions in the Atlantic and the Channel. British forces did, however, continue to participate in exercises and periodically contributed to NATO's on-call force when present in the Mediterranean. The Alliance also benefited from the continuation of British base rights on Cyprus and Malta (AWEU 1973: 8; 1976: 15).

Allied use of the British facilities on Malta was put into question with the election of the Mintoff government in 1971. NATO moved COMNAVSOUTH to Naples. However, in March 1972, Malta signed an agreement with the UK by which Britain secured the right in peace and war to station its armed forces in Malta 'and to use facilities there for the defense purposes of the United Kingdom and of the North Atlantic Treaty Organization'. Under Article 2 of the treaty, the government of Malta pledged not to 'permit the forces of any party to the Warsaw Pact to be stationed or to use military facilities there' apart from harbor facilities (AWEU 1972: 10, 17). It would appear that NATO and particularly elements of the Sixth Fleet did not, in subsequent years, avail themselves of the facilities on Malta, although Soviet ships did dock there.[36]

Indeed the removal of COMNAVSOUTH to Naples allowed for better integration of NATO's command near CINCSOUTH headquarters and the submarine command, CONSUBMED. The US Sixth Fleet maintained a liaison office at Naples headed by the second in command responsible for NATO matters.[37]

Also in Naples was the headquarters of Maritime Air Mediterranean (MARAIRMED). Subordinate to COMNAVSOUTH, this new command, created by the Defense Planning Committee in November 1968, brought together the land-based maritime air units of the US, Britain, Italy, Greece and Turkey. By placing maritime air under the control of the commander (COMNAVSOUTH) who controlled the surface and submarine units, it was hoped that the Alliance would be better able to control the sea lines of communication. In peacetime, MARAIRMED provided a basis for co-ordination of day-to-day surveillance and reconnaissance (see Marriott 1968).

In April 1969, the DPC authorized the creation of a naval on-call force in the Mediterranean (NAVOCFORMED) under COMNAVSOUTH. First activated in April 1970, NAVOCFORMED was something between the STANAVFORLANT and MCFs in the Atlantic in that it consisted of a multinational squadron of 3 to 5 destroyers called together periodically. As with the standing force in the Atlantic, this new NATO force was meant primarily as a means of demonstrating allied solidarity and to further training and inter-naval familiarization with NATO operating procedures.[38]

Thus, although NATO could no longer be assured of immediate and decisive control of the Mediterranean the Alliance had taken a number of steps to enhance the readiness and effectiveness of its maritime forces along the southern flank. Commenting on the situation in 1972, the then Sixth Fleet Commander Isaac Kidd noted: 'NATO works in the Mediterranean. It is working daily in the invaluable exchange of information and perfecting of operating techniques at sea on bilateral and multilateral exercises' (Kidd 1972: 29).

The emphasis on improving NATO's peacetime posture at sea so as to better support a strategy of flexible response was also apparent in Channel Command. Here the Alliance wanted to be able to keep crucial ports open to allow for reinforcement and resupply and for continued civilian imports. Thus, in May 1973, a Standing Naval Force Channel (STANAVFORCHAN) was commissioned. Operating permanently under CINCHAN, the force consisted of nine counter-measure vessels from Belgium, Denmark, Germany and the UK. American and Norwegian

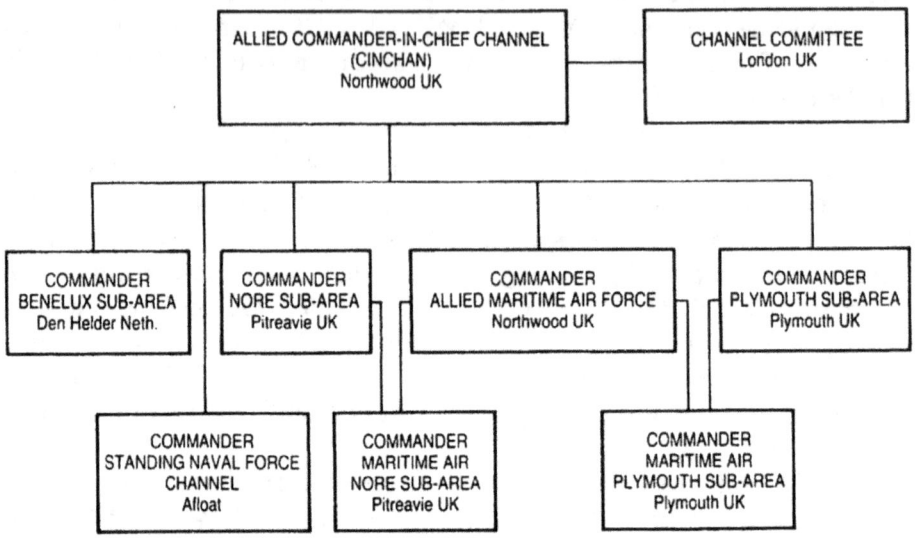

Figure 4.3 Allied Command Channel

ships would also join STANAVFORCHAN from time to time (NATO 1981b: 140).

THE BROSIO STUDY

The changes in NATO's maritime command structure, and the emphasis on peacetime measures to improve the co-ordination and responsiveness of allied maritime forces, indicated the extent to which allied naval leaders, SACLANT in particular, had succeeded in conveying the importance of the seas to the Alliance's political leadership. On 24 January 1968 SACLANT had presented to the North Atlantic Council his appreciation of the 'World Wide Soviet Maritime Threat'. As a result, the Secretary General, Manlio Brosio, directed that a study be undertaken on NATO's maritime posture. The Chairman of the Military Committee then asked SACLANT to undertake this study; he, in turn, selected a group of naval officers from the various allied navies represented on his staff to examine: (a) the relative strength of the maritime forces of NATO and the Soviet Bloc, worldwide; and (b) the maritime strategic doctrines of the two sides.[39]

The *Brosio Study* was presented to the Secretary General on 19 March 1969. It was a wide-ranging study which examined not only the respective strengths of the two maritime forces, but also the economic and political contexts in which each was developed. Included as well was an evaluation of one particular conflict scenario, a limited conflict along the northern flank. This scenario was run for the forces existing at the time of the study and run again using estimated forces for a 1977 timeframe.

The *Brosio Study* set forth NATO's maritime strategy in light of the growth of the Soviet Navy and the Alliance's new strategic doctrine of flexible response. Flexible response, it was stressed, expanded the contribution of maritime forces to NATO's overall deterrent posture and war fighting capability. Support of the land and air forces ashore and protection of the sea lines of communication continued to be the primary tasks of allied maritime forces. Now, however, NATO would look to its forces at sea to fulfill these tasks throughout the entire conflict spectrum, from normal peacetime conditions, through limited war, to major conventional and nuclear war. The study emphasized the importance of early pre-hostilities deployment of maritime forces and the need to be able to respond to Soviet actions at sea under conditions of limited war. To the overall strategy of flexible response, the allied maritime forces would bring the flexibility and mobility of maritime forces.

In normal peacetime conditions, the *Brosio Study* pointed out that NATO had to ensure that its maritime forces contributed to the Alliance's conventional and nuclear deterrent. These forces would demonstrate the capability and unity of NATO such as to counter any Soviet maritime activities directed toward gaining influences at the expense of the West. The newly created Standing Naval Force Atlantic would contribute by providing training facilities, joint surveillance and an initial collective presence in the event of tensions with the Soviet Bloc.

If tension continued to mount, with perhaps some Soviet limited acts against NATO forces or shipping, the Maritime Contingency Forces would be mobilized. The purpose of the MCFs would be to provide a controlled response of sufficient size and strength to compel the Soviets to choose between withdrawal and escalation. The contingency forces would also seek to deter, and in the case of failure of deterrence, prevent the Soviets from achieving any quick, limited *fait accompli*. In limited war situations, the *Brosio Study* suggested that in support of NATO's flexible strategy maritime forces might respond to Soviet pressures on land or in the air by applying pressures at sea against the Soviet Bloc as a deterrent against further escalation.

In the event that a limited conflict evolved into a major war, allied maritime forces would undertake a full range of actions according to the responses selected by NATO's political leadership. In general, the study noted that under conditions of major war, the Alliance would seek to maintain or re-establish overall supremacy at sea and employ its maritime forces in various capacities to support the land and air battle. This would encompass operations to contain and, if necessary, destroy Soviet submarines as far forward as possible; the use of strike task forces to support land battles with additional aircraft and amphibious forces, especially along the flanks; mining and countermine measures; naval control of shipping and protection of merchant shipping subsequent to, or in concert with, offensive operations against Soviet forces (depending upon the availability of forces); and theater and strategic nuclear strikes launched from carrier strike fleets or SSBNs.

While the *Brosio Study* evaluated the global capabilities of the Soviet and NATO maritime forces, it is evident that the members of the study group saw the immediate European waters as the location for major Soviet and allied maritime deployments in the event of hostilities. NATO forces would begin a build-up in these waters in advance of any conflict in support of the flexible response strategy. The tasks for Soviet forces would be to counter this build-up and, should conflict break out, destroy NATO maritime forces, particularly those supporting in any way the land and air battle ashore. In addition, the Soviets would seek to maintain control of strategically important sea areas adjacent to the Soviet Union and, if necessary, attack land targets from the sea, perhaps with SLBMs fired from the protected seas.

It was this kind of encounter at sea – NATO moving in to support forces ashore with strike aircraft carriers, amphibious ships, merchant marines carrying reinforcement, all protected by ASW forces opposed by the Soviets – upon which the *Brosio Study* based its general conclusions. It said that, as of 1968, the Alliance could expect to secure control of the seas against Soviet forces in a major conventional maritime struggle. The struggle could take over a month and there would be major losses on both sides, but in the end NATO would prevail at sea.

Extrapolating the then current trends in Soviet and NATO maritime developments, the study examined the outcome of the sea war in 1977. It concluded that by the mid to late 1970s, the Alliance would no longer be assured of being able to maintain control of vital sea areas without a major naval building program on the part of the allies. Moreover, the expected growth in Soviet capabilities would not entirely be offset by the NATO navies since only a portion of allied naval forces were earmarked for

Alliance use; and even if, in the event of war, more were made available, it would take some time for these forces to be brought up to war readiness. The study also stressed the dangers of the Soviet Union's wide use of surface-to-surface and air-to-surface missiles.

The study contained an in-depth operational analysis of only one scenario: a limited war along the northern flank. This scenario was selected not so much because this is how the study group believed a war might occur, but rather it afforded a maximum interplay of conventional maritime capabilities.

In this scenario, war began with a Soviet invasion of northern Norway using amphibious and airborne forces, while naval units and other forces were mobilized for conventional and nuclear war. NATO had several days of warning during which time carrier task forces (the number varied according to different subscenarios) proceeded to the Norwegian Sea. Other carriers accompanied by destroyers and support ships moved into the Atlantic from American ports, while additional destroyers were mobilized for convoy escort duties. Anti-submarine barriers were deployed and maritime patrol aircraft began flights over the GUIK gap. Fighter aircraft squadron were deployed to counter Soviet long-range naval aviation.

On D-day, carrier-based aircraft attacked the Soviet invading force and were simultaneously attacked by Soviet naval aviation while NATO began concerted ASW against Soviet submarines. The Norwegian Sea became a battleground for the opposing naval forces with both sides using only conventional weapons. Soviet SSBNs and nuclear armed SSGNs were at sea, some stationed off the US and the UK to deter against nuclear use by NATO. Allied shipping was subjected to attack and convoys were organized. The war lasted 60 days, after which time the Soviets launched a major offensive across the entire central front, having failed to achieve their limited objectives in Norway.

Using only earmarked forces, NATO was able to defeat the Soviets at sea for the 1968 scenario. The carrier forces could handle Soviet air attacks while supporting the land battle. The more carriers available, the more effective they were against Soviet air attacks and in terms of projecting force ashore. Additional carriers allowed NATO to rotate them above and below the GUIK gap for replenishment out of range of Soviet forces.

The study also concluded that NATO ASW efforts could contain the Soviet submarine threat and that the Soviets would use most of their submarines for defensive purposes with few venturing out for anti-shipping attacks. The Alliance could afford to allocate relatively few destroyers for convoy escort, and during the 60-day war were able to keep merchant ship sinkings reasonably low. (The study did not test for special military convoys from North America carrying SACEUR's strategic reserve but did note that these would be the most profitable targets for the Soviets and therefore they might take more chances trying to sink them.)

The identical scenario run for 1977 ended in NATO failing to control the seas. This was due in large part to what was anticipated to be a continued rapid growth in Soviet naval forces and enhanced Soviet capability for underway replenishment. With additional submarines at sea, ASW barriers were unable to prevent enough submarines from reaching the sea lanes and merchant shipping losses tripled. Escort forces could not contain the submarine and air threat.

As with all hypothetical scenarios and war gaming, the results contained in the *Brosio Study*'s operational analysis depended on the assumptions made. For example, only earmarked maritime forces were counted and the French Navy was not included in the allied order of battle. This reduced available NATO forces. On the other hand, the assumption that several carrier task forces would be available to support allied forces on the ground and in the air in Norway may have been too generous. The very fact that there was no other threat, either inside or outside the NATO area, may also have been unduly optimistic. NATO might have lost the war at sea in 1968, had the Soviets employed nuclear armed SSM, but allied ASW measures may have been more effective if nuclear ASW weapons had been employed. Another consideration is the course of the land battle. In the 1968 victory, it took several weeks to secure the Norwegian Sea, during which time the land front had held.

NATO MARITIME DEVELOPMENT IN THE 1970S

What the *Brosio Study* did affirm for the first time in a NATO assessment was that the Alliance would have to wage a major campaign at sea if it was going to be able to exploit the strategic value of the seas for reinforcement and resupply and for projection of force ashore. In the 'comprehensive' study, *Allied Defence in the Seventies* (AD-70), presented to the Defense Planning Committee in December 1970, attention was directed to the maritime threat facing NATO. The DPC called for 'better maritime surveillance and anti-submarine forces, more maritime patrol aircraft and seaborne missile systems, and the replacement of overage ships' (NATO 1982: 271).

The US Navy was particularly anxious to have its allies contribute more to NATO's maritime posture. The number of ships in the American fleet had been declining and although the newer vessels were more potent than the ones replaced, the USN viewed its global commitments as stretching it capabilities in the event of war. It might not be possible for the USN to meet its NATO and other responsibilities simultaneously.

Thus, in 1970, CNO Elmo Zumwalt asked the US Naval War College to develop a plan designed to improve the ASW capabilities of the NATO allies. The *Newport Study* was undertaken with the assistance of US naval commanders then associated with NATO naval commands.[40] It was completed in April 1971.

In a letter to Zumwalt dated September 1971, Admiral Colbert (by then SACLANT Chief of Staff) reported that he was working with Admiral Rivero (CINCSOUTH) on implementing the recommendations of the *Newport Study*. He wanted the US Defense Department to support their efforts by persuading the allies to 'carry a heavier burden at sea'. Colbert called for the 'elimination' of the alleged McNamara policy of 'downgrading NATO navies'. It was also important to establish with the US Joint Chiefs of Staff 'the fact that the shift in the balance of the threat has been to seaward and that MAP [Military Assistance Program] allocations to strengthen allied naval forces need to reflect this'.[41]

Throughout the 1970s, the total number of allied maritime forces declined. However, it would appear that the concerns raised by NATO naval leaders did help to persuade the NATO governments to upgrade their maritime forces. The Europeans and Canadians continued and intensified their traditional specialization in the areas of ASW, mine warfare and coastal defense.[42] Some navies, such as those of the British and Dutch, reduced their blue water capability, the Royal Navy in particular reducing its attack carrier force.

The Royal Navy was also pulling back 'East of Suez'. This withdrawal was mainly for economic reasons. But, as Eric Grove has argued, 'the improvements in Soviet naval capability (not so much in quantity of submarines and ships but in their quality and level of deployment) made a move westwards well advised' (Grove 1987: 305). The British government thus used the adoption of flexible response with its emphasis upon conventional defense in Europe to bolster the case for overseas withdrawal and simultaneously to provide a revitalized broad strategic rationale for the naval building program:

> our withdrawal from overseas will enable us to increase the number of ships at immediate readiness for NATO's shield forces, and so enable us to continue to play a leading part among the European navies in the NATO maritime alliance ... the growth of Soviet maritime strength ...

has underlined the importance of shield forces especially in relation to the flanks of Europe, Scandinavia and the Mediterranean, where the increase in Soviet naval presence has been most evident.[43]

Nearly all allies began major replacement or upgrading programs. For example, Belgium commissioned four Wielingen Class frigates, significantly improving its contribution toward protection of shipping in the Channel. Germany began to replace its 1950s era frigates with the German-built Bremen Class frigate and its submarine fleet with the German-built 'type 210' submarine, as well as a new generation of fast patrol boats with modern vessels carrying the American Harpoon missile (Roberts 1981: 35). Overall, the Alliance's navies moved to increase their surface firepower through deployment of surface-to-surface missiles such as the Penguin and Exocet (Underwood 1979: 15–16). A number of navies acquired additional long-range aircraft to augment ASW capabilities.

The improvements made in the capabilities of the allied navies during the 1970s meant that, as one American Admiral told a Congressional hearing, non-US units still supplied 'substantial maritime forces for the common defense'. Nearly 65 per cent of all 'ocean-going' NATO maritime forces and 26 per cent of major combatants would be drawn from Europe and Canada (US Senate 1978: 1737). The US Navy continued to supply NATO with essential protection and projection forces, including the seaborne nuclear deterrent. But in terms of conventional forces in anti-submarine warfare,[44] mine warfare and coastal defenses, other allies would play an important role. This would be the case especially in the initial stages of any conflict where the European forces would bear the brunt of maritime tasks in the eastern Atlantic (Longworth 1983: 88). Indeed, while SACLANT remained the most important NATO maritime command, by virtue of inclusion of the US Atlantic fleet, the allied headquarters at Northwood near London would be a vital point for command of the war at sea. There the staffs of CINCEASTLANT and CINCHAN would direct allied forces in the North and Norwegian Seas and in the Channel.[45]

The improvement in the maritime capabilities by most allies took place amid enhanced efforts by the Alliance's political and military leadership to strengthen NATO's maritime posture. In 1976, SACLANT and CINCHAN undertook a series of maritime studies, concurrent with SACEUR's flexibility studies which resulted in an extensive prioritized list of recommended short-term improvements in the allied posture at sea.[46]

The Long Term Defense Program (LTDP) set forth by the Defense Planning Committee in May 1977 and adopted by the NATO Council in Ministerial session a year later stressed the importance of improving the Alliance's maritime posture. In particular, the LTDP called for increased allied co-operation in the development and production of weapons, platforms and advanced technologies related to naval warfare. These included surface-to-surface and surface-to-air missiles, missile defense systems, communications, data links and mine warfare capabilities. At the May 1978 DPC meeting, attention was called to the continuing installation of medium-range active sonars now carried by most of the Alliance's surface ships, to the planned increase in the stocks of lightweight torpedoes capable of being dropped from aircraft (i.e. the US MK-46 Mod 5 Neartip) and the efforts of the NATO Naval Armaments Group (NNAG) to develop a new lightweight torpedo and an advanced acoustic sensor (North Atlantic Assembly 1982: 32).

The LTDP sub-area on maritime command, control and communications addressed the further implementation of Link 11, 'a standardized NATO automated computer to computer digital data link to convey tactical information between maritime forces'. By 1980, Belgium, Canada, France, Italy, the Netherlands, Norway, the UK, the US and West Germany had already installed, or were planning to install, Link 11 in surface ships. Some had also put the system into maritime patrol aircraft. Another area of NATO efforts at interoperability was sonobuoys. In 1978, the French Navy hosted a demonstration which indicated a 98 per cent interoperability of American, French, West German, Canadian and British sonobuoys with US, French, British and Canadian maritime patrol aircraft (ibid.: 32–3).

Co-operation was extending to major vessels such as the jointly produced Dutch–German frigate used in the navies of those two countries as well as by Greece. The Alliance was looking into a follow-on to this type which might be used to replace ASW frigates in the 1990s. A number of countries had co-operated in the development of the NATO Patrol Hydrofoil Missile Ship (PHMO), and the US and Canada began testing a Canadian-designed variable depth sonar which could be towed by high-speed ships like the PHM (ibid.). One joint approach to missile defense was the NATO Sea Gnat project. This was a co-operative effort by the US, UK, West Germany, Norway and Denmark to develop a decoy system. A major project designed to enhance allied surveillance was the Azores Fixed Acoustic Range (AFAR), which involved implanting transmitting and receiving antennae on the ocean floor. The range was active between May 1972 and May 1979 (NATO 1981b: 172). With funding from the NATO infrastructure program, the Alliance developed two Naval Forces Sensor and Weapons Accuracy Check Sites (FORACS), one off Norway and the other near Crete. The object here was to 'acquire, equip and operate ranges for the purpose of checking the accuracy of all sensors connected with shipborne weapons systems' (ibid.: 169).

Efforts to standardize or at least to increase the compatibility of platforms and weapons used by the allies were accompanied by an expansion of efforts to standardize procedure and tactical doctrine. Common manuals were being used by the NATO navies. In 1980, the major NATO commands (SACEUR, SACLANT, CINCHAN) completed a joint study which produced a formalized, agreed 'concept of maritime operations' (CONMAROPS). The development of CONMAROPS reflected the new importance now being attached to NATO's maritime forces. But it was not a radical departure from previous concepts. Rather, as Eric Grove has observed, it was 'a re-articulation in a much more confident and coherent form of ideas that have been around ever since Atlantic Command was created in 1952' (Grove 1989: 3). It would be the core document for NATO maritime forces. According to SACLANT, Admiral Harry Train II, USN, the concept:

> identifies NATO maritime interests and assesses threats to these interests, considers the type of confrontations that can be expected and the associated allied priorities; establishes the principles to be used by NATO forces, and finally outlines the campaigns that are likely to be waged and the involvement of various types of forces. The three principles are – containment, defense in depth and keeping the initiative.
>
> (Train 1982: 24)

With its emphasis upon containing the Soviet Navy in the forward areas and in gaining and maintaining the initiative, CONMAROPS can also be viewed as a prelude to the United States Navy's much heralded and much maligned 'maritime strategy' of the 1980s. Where they differed was not so much on the need for forward maritime operations, but where and for what purposes. For CONMAROPS, the 'overriding requirement' was 'to support the Supreme Allied Commander ... in Europe' especially through the 'protection of military reinforcement sealift and resupply' (NATO 1981a: 14). The USN's 'maritime strategy' placed great emphasis on this too, as indeed it would have to since SACLANT was still Commander-in-Chief, Atlantic, for the United States. But the 'maritime strategy' called for forward operations outside the NATO area, horizontal escalation and placed particular emphasis upon forward operations against Soviet

SSBNs in their northern bastions (see Watkins 1986). It would have been difficult at best to get allied governments formally to go along with this, nor was there any need since the USN was prepared (and preferred) to act unilaterally in these roles. But as far as maritime forward defense in support of SACEUR went, CONMAROPS was consistent with the USN's 'maritime strategy' and both represented a continuation of the approach taken by the American and other allied navies to the defense of Europe since 1949.

The development of standardized procedures and doctrine was furthered by real-life exercises. During the 1970s, allied maritime exercises became larger and more complex with an emphasis on early mobilization and controlled flexible responses, particularly along the flanks. Exercise STRONG EXPRESS, conducted in September 1972, was the largest ever combined exercise involving some 300 ships, 700 aircraft and 64,000 personnel from 12 nations. It tested amphibious landings, projection of air force ashore in support of the northern flank operations, convoy protection and ASW in the Norwegian Sea, North Sea and English Channel.[47] That same year, NAVSOUTH held DAWN PATROL in the Mediterranean, involving 80 ships and 300 aircraft, including those attached to MARAIRMED. There was activity throughout the Mediterranean aimed at countering 'enemy' surface, subsurface and air forces, culminating with an amphibious landing in southern Greece supported by a task force which included British and American carriers, Greek minesweepers and Italian and Turkish destroyers.[48]

In exercise OCEAN SAFARI 77, which took place in October 1977, NATO tested 'command, control, communications and tactics with the theoretical use of nuclear weapons while an imaginary war escalated on the central front'. A reinforcement convoy on its way to Europe amid a deteriorating political situation was first shadowed and then attacked once general hostilities began. Until the initiation of hostilities by the 'enemy', NATO forces had to avoid any provocation 'or aggressive act'. Combat, once begun, included use of nuclear weapons against allied ships at sea (Canada, Dept of Defense 1977).

Also receiving more attention in the 1970s were allied procedures regarding reinforcement. In 1980 and 1981, the major NATO commanders completed a series of reinforcement studies directed toward identifying measures for improving airlift, sealift and other aspects of reinforcement. One result was the creation of a specially earmarked pool of allied merchant ships which would be available for the sealift of American forces early in a crisis.[49] (Improvements in the availability of reinforcement shipping are detailed in the following chapter.) In exercises held in the late 1970s, such as the REFORGER and WINTEX series and NIFTY NUGGET, the US and NATO tested sealift mobilization capability as it had not been done in the past (US Dept of Defense 1978).

Allied naval leaders had always stressed the global nature of the Soviet maritime threat, particularly to Europe's oil lines of communications from the Persian Gulf. NATO's political leadership was reluctant to extend the Alliance's sphere of concern, even below the Tropic of Cancer. This did not mean that several allies, acting unilaterally or in co-operation with one or two other nations, could not co-operate out of the area. During the 1970s, the US, Britain and France did begin joint exercises and other forms of collaboration in the Indian Ocean. Several NATO steps were, however, taken in relation to out of area threats; SACLANT was authorized to be prepared to protect shipping below the Tropic of Cancer as necessary and to proceed with contingency planning in this regard. At Channel Command/EASTLANT headquarters a Maritime Exercise Information Group was established to collect and distribute information on the movements of allied navies throughout the world (Swartztrauber 1979: 116).

Although the main thrust of flexible response was to enhance NATO's ability to fight a conventional war and raise the nuclear threshold, it was also, as noted, intended to provide the Alliance with a more discriminating approach to nuclear war. In the 1970s, allied maritime forces improved their capacity to contribute to theater nuclear forces. By the end of the decade the US had some 400 MIRVed Poseidon SLBMs allocated to SACEUR. These were believed to be targeted against, among other points, Soviet medium missiles aimed at Western Europe (Sorrels 1983: 72). The British and the French were also moving to improve their sea-based ballistic missiles. In 1979, the Thatcher government decided to replace their Polaris SLBMs dedicated to NATO with 100 American Trident Is to be deployed in four new SSBNs. The French government intended to introduce the M-4 MIRVed SLBM (ibid.: 52, 68).

By 1977, the American Sea Launched Cruise Missile (SCLM) was being discussed as one of the options for a new long-range theater nuclear delivery system. With its 1,550 to 2,500 miles operational range, the Tomahawk SLCM could hit targets in the western Soviet Union from ships stationed in the Norwegian and North Seas (ibid.: 80–1). Although NATO, in its December 1979 decision, selected new land-based systems, the deployment of SLCMs on US submarines and surface ships would greatly augment NATO's *de facto* theater capability.

While American strike carriers had become increasingly vulnerable to Soviet naval aviation, additional carrier-based tanker support allowed the A-6s and A-7s to operate at combat radii in excess of 1,000 kilometers.

This provided extra depth in which to counter Soviet missile-firing bombers. Another improvement was the deployment of the AEGIS defense system which provided an enhanced ability to identify, track and intercept surface-to-surface and air-to-surface anti-ship missiles.

The development of missile defense systems, SLCMs, which could hit ships as well as shore targets, and such undersea nuclear weapons as the B-57 air-dropped nuclear depth bomb, seemed to increase NATO's ability to conduct a tactical nuclear war at sea. This capability would be enhanced if early maritime combat did not involve extensive protection of shipping but rather quickly escalated into an East–West test of wills on the oceans. A more complex scenario would be the use of nuclear weapons to destroy allied ports, thus preventing reinforcements and NATO responding with a wider use of nuclear weapons against Soviet maritime targets. Whatever the possible circumstances, by the early 1980s, the US Navy appeared to be giving more consideration to a 'revitalization' of tactical naval nuclear capabilities of a kind that had long been germane to the NATO maritime posture (Arkin 1983: 7).

NATO IN THE SEVENTIES: THE MARITIME IMPERATIVE

In the largest sense, NATO had not changed its maritime strategy during the 1970s. That strategy remained one of seeking to secure the vital seas in order to allow for conveyance and projection of military force. What the Alliance had done was to give greater attention to the means by which the strategic value of the seas would be exploited in the event of a land/air war on the continent – a war of indeterminate duration and uncertain character. The twin developments of a growth in Soviet maritime capabilities, coupled with the adoption by NATO of a flexible response strategy, had increased the demands on allied maritime forces in the context of its overall deterrent posture.

With the growing importance of conventional deterrence within that overall posture, the inherent maritime nature of NATO had been brought to the fore. To be sure, allied, mainly American maritime, forces had long contributed to NATO's nuclear deterrent posture. In the 1970s this contribution was enhanced with the development of a more accurate MIRVed SLBM and sea-launched cruise missiles. But it is evident that the major impetus for the greater demands being placed upon NATO maritime forces, and hence the greater emphasis upon them in the 1970s, was concern over the Alliance's conventional posture. As in older calculations regarding the maintenance of the balance of power in Europe, account had to be taken of the ability to sustain land and air forces in war.

The collective maritime forces of NATO had come to represent the pre-eminent example of seapower in the nuclear age. Given the catastrophic consequences that could flow from a failure to maintain an adequate conventional war-fighting capability, NATO's maritime forces had, moreover, assumed a burden that maritime forces had never had in the past. NATO had always been a maritime alliance; now it seemed imperative that it remain so.

Chapter 5
Reinforcement sealift

SEALIFT AND NATO POSTURE

This study argues that the role of seapower in the nuclear age remained fundamentally the same as in previous times – that is, seapower was directed towards utilizing the strategic value of the ocean space, in particular the access afforded by the seas to non-adjacent land areas. Thus, the ultimate objective of any maritime strategy was to support the exercise of military power ashore. The most common strategic use made of the seas in support of land operations has been the conveyance of troops and equipment to battle areas. Alfred Thayer Mahan's classic work on seapower opens by comparing Rome's 'control of the water' with Hannibal's inability to draw support by way of the sea. It was this difference, according to Mahan, which foretold the victory of the former and the defeat of the latter (Mahan 1957: xii).

The present chapter examines NATO's efforts to assure adequate sealift for the movement of American forces and their supplies, as well as military supplies for other allied forces, to Europe in the event of mobilization or war with Warsaw Pact forces. Few other aspects of NATO's maritime strategy have received as much criticism as plans to reinforce the land and air forces in Europe by sealift. The main objections have centered around unrealistic expectations as to the length of a conflict, in particular the duration of a conventional struggle given the presence of tactical and strategic nuclear weapons. Any war in Europe would be too short or too destructive for sealift to make a difference. A further commonly heard complaint is that the Alliance never had an effective peacetime plan or organization to assure that sufficient numbers of the right kind of vessels could be available on time.[1] Thus, efforts on the part of the allies, especially those of the NATO navies, to assure adequate, timely and useful sealift capability, have been portrayed as somewhat irrational responses to the existing strategic situation. It was a case, once again, of military men planning for the last war, in this instance drawing upon the lessons of combined allied shipping operations during the Second World War.

If NATO's military and political leaders have from the outset sought to ensure a measure of sealift capability and Alliance co-operation in supplying that capability, it was because of the continuing importance of land power in allied strategic calculations. This, in turn, was a reflection of uncertain expectations as to the nature and duration of a conflict with the Soviets and their allies. The very creation of a combined command structure, and the deployment of large standing forces in peacetime, indicated a commitment to some measure of conventional territorial defense commensurate with the principal aim of the Alliance. As Nils Ørvik observed in the early 1960s, that aim was 'to protect the inviolability of NATO territory' (Ørvik 1963: 6). As long as even a limited conventional territorial defense was possible, even one combined with the use of tactical nuclear weapons, there was a perceived need to assure reinforcement and resupply of the NATO, principally American, forces.

According to a 1956 article by the British Admiral Sir Michael Denny, the eastbound shipping across the Atlantic in the event of a NATO war, 'should be properly regarded as the lifeline' the 'safeguarding of which and its provisioning by an adequate flow of merchant ships is an important prime factor in the NATO setup'. Not even the increased importance of airpower had reduced the need for sealift. 'On the contrary,' argued Denny, 'the heavy demands of air forces for operational maintenance and for fuel have greatly increased demands on seaborne transit' (Denny 1965: 352–4). Though he expressed doubts about the adequacy of the existing plans, he noted that NATO's 'political and military leaders take it for granted that the seas will be secure' (ibid.: 358).

Assessments of US sealift requirements and capabilities generally reflected overall Alliance expectations as to when and how the American reinforcement and resupply effort would come into play. A 1962 Rand Corporation study on US military shipping requirements noted that the existing programs to ensure availability were 'geared toward a war short of all-out nuclear war'. The only circumstances under which the United States would find itself short of sufficient shipping would be a war 'many times the size of the Korean War'. At present, the study argued, Europe was the only place where 'a great augmentation' of forces would be required such as would go beyond the limits of US sealift capabilities. In this instance, however, the American forces would 'have the use of the world's largest block of vessels and seamen, those of the Western European countries' (Rapping 1963: 24, 42).

The expectation that not only would the US need the additional sealift capacity supplied by its NATO allies, but that this additional capacity would be forthcoming, was the basic consideration that informed Alliance planning in this area. To be sure, the US Navy, with the support of the American merchant marine and shipbuilding industry, continually sought to increase American flag shipping on national security grounds, to the point of casting doubt on the credibility of allied shipping commitments. Yet, throughout the period with which this study deals, and more particularly in the latter years with the decline in American flag shipping, it was a central tenet of allied maritime strategy that a US reinforcement effort would require NATO-wide co-operation in the provision of sealift capacity.

This co-operation was needed not merely to ensure adequate capacity. Taken together, the NATO allies had more than enough shipping to meet American reinforcement needs. For the US, the sealift problem was not so much a matter of total capacity as early availability.[2] To be an effective supplement to United States flag shipping, NATO vessels had to be at the disposal of the US military as quickly as possible.

THE EARLY YEARS: PBOS AND THE DSA

At its May 1950 meeting, the North Atlantic Council established the Planning Board for Ocean Shipping (PBOS). PBOS was to report directly to the Council and 'work in close co-operation with other bodies of the Treaty Organization in all matters relating to the factor of merchant shipping in defence planning' (NATO 1982: 55). Eventually, the Board became a sub-agency of the Council's Senior Civil Emergency Planning Committee (SCEPC).

Although a NATO body open to representatives from all member nations, PBOS was to all intents and purposes an Anglo-American operation. Even after the NATO Council moved to Paris, it continued to be operated from London and Washington. As Secretary-General Lord Ismay noted in 1955, PBOS, unlike other NATO agencies, was not served by an international secretariat nor were its operations supported by common funds, rather: 'Its work is carried on chiefly by a small group of British and American civil servants who, with the co-operation of shipping experts from other member governments, prepare papers for the Planning Board's consideration' (NATO 1955: 145).

At its initial meetings, the Board agreed that in order to 'diminish the effects of a shortage of sea transport at the outbreak of war, the great bulk of ocean-going merchant ships under the flags of NATO countries would be pooled' and placed at the disposal of an inter-allied Defence Shipping

Authority (DSA). The DSA would consist of a Defence Shipping Council and a Defence Shipping Executive Board (DSEB), with the Council formulating general shipping policy, 'in accordance with the overall strategy', and the DSEB administering the central pool 'and for the purposes of day-to-day operation' operating with two branches, one in Washington and the other in London. While actual operations would not begin until the outbreak of war, it was important to ensure that both branches could be brought into being early enough to be 'fully operative' when hostilities commenced. 'To this end,' notes Ismay, 'steps have been taken for assembling the key personnel, including national shipping representatives of member countries' (ibid.: 146). In addition, the allied governments had prepared lists for PBOS showing the characteristics of their fleets, some of which would be allocated to the central pool to be administered by the DSEB. From these lists, PBOS, which met yearly (alternately in Washington and London), began to compile combined lists of available allied shipping to meet NATO requirements. Tables 5.1 and 5.2 are examples of PBOS studies in 1952.

Table 5.1 Eastern hemisphere dry cargo merchant fleets. Estimated ocean cargo tonnage available on average throughout the first year of a war based on 1949 and 1950 total tonnage figures

	Vessels of 1600 gross tons and over		*Million deadweight tons*	
	31 December 1949		*31 December 1950*	
Estimated total tonnage		35.5		37.0
Deduct losses in year (average 6%)		2.1		2.2
		33.4		34.8
Add new building in year (average 50% of completions)		0.6		(say) 0.6
		34.0		35.4
Deduct				
Passenger tonnage	4.3		4.2	
Coasting and short sea tonnage	5.7	10.0	6.1	10.3
		24.0		25.1
Deduct immobilized by repairs (12½%)		3.0		3.1
Estimated ocean cargo tonnage available throughout year		21.0		22.0

Source: Records of the United States Maritime Administration, Planning Board for Ocean Shipping, Document PBOS/4/6, 9 May, 1952.

Table 5.2 Particulars of ships with heavy derricks as at 31 December 1950 – summary of Eastern hemisphere NATO countries (excluding Greece and Turkey)

Derricks safe working load of heaviest derrick carried long tons (2240 lb)	Number of ships carrying derricks as in column (a)			
	Ordinary Merchant ship types			Specialized heavy lift ships e.g. BELS, CLANS
	Refrigerated	Liners	Tramps	
(a)	(b)	(c)	(d)	(e)
Over 120	–	–	–	6
100–120	–	–	–	15
80–99	7	19	21	–
60–79	8	17	2	–
50–59	51	290	416	–
Total	66	326	439	21

Notes: 1. Two of the countries included in the above summary have classified the safe working load of derricks in metric tons, but, for the purposes of this summary, classification by metric tons has been assumed to be the same as by long tons.
2. Passenger vessels are excluded.
3. For this summary, a refrigerated ship is one with 40,000 cubic feet or more of refrigerated cargo space.
Source: Records of the United States Maritime Administration, Planning Board for Ocean Shipping, Document PBOS/4/5, 9 May 1952.

When it became operational, each branch of the DSEB would establish several committees: shipping deployment, allocation, tanker committee and freight rate committees. Together, the DSEB and its committees would relate the total demands upon allied shipping 'as put forward by the various claimants for overall available tonnage'. Demands would come from both civilian and military authorities. According to a memorandum prepared by the Joint Military Transportation Committee of the US Joint Chiefs of Staff, the allocation of shipping would be made 'in accordance with priorities laid down by the central authority charged with the higher direction of the common effort'.[3]

The major claimants on allied shipping would be the military leaders, in particular the American and British commanders in charge of reinforcement and resupply. It is significant that PBOS was established prior to the creation of the integrated military command structure and prior to the Korean War-inspired build-up of NATO's standing forces. Combined with its essentially Anglo-American character, this indicates an early allied concern for adequate and timely sealift capacity to bring US and

UK forces to Europe as quickly as possible. In addition, these early plans appear to confirm that allied expectations as to the character of a future war in Europe certainly did not rule out a protracted struggle. It would be one in which land forces would be the principal means by which the Soviets were driven back, strategic nuclear bombing notwithstanding.

With the establishment of the position of SACEUR and a Supreme Headquarters of Allied Powers Europe (SHAPE) to oversee the standing forces of the Alliance, more specific steps were taken to measure the requirements and availability for a reinforced sealift. The creation of a NATO standing conventional force, therefore, did not lessen military concern for assured additional support in the event of mobilization. Indeed, it can be argued that once the US in particular (as well as Canada) committed forces to forward defense, sealift support followed automatically since no military commander would place an army (with air forces) in the field and neglect logistics. In this case, the rear areas of the allied armies extended back across the Atlantic to the US and Canada and across the Channel to England. Whether the American, Canadian and British forces were a 'shield' or 'tripwire', or just a political demonstration of allied solidarity, the traditional military necessity for adequate reinforcement dictated sealift preparation. Moreover, given the state of the European forces, there would be a need to resupply them if they were to sustain resistance along with the expeditionary forces.

In November 1951, the United States Joint Chiefs of Staff received a request from the US representative to the standing group of the North Atlantic Council's Military Committee regarding SHAPE's shipping requirements. The request asked for information on the availability of shipping to meet US requirements, the length of time required to deploy US and Canadian forces and 'the increase in the number of notional[4] ships required for US forces during the first year of the war'.[5]

The JCS reply gives a fair indication of the considerations involved in planning for a sealift of US forces in support of SACEUR (it did not deal with Canadian forces) and of the state of NATO plans to provide sealift capacity. It was noted, first of all, that no exact answer could be given as to the availability of shipping to meet SHAPE's requirements, 'because a pre-allocation of shipping from the NATO pool has not been made'. On the basis of the number of ships in the US inventory (US private and government-owned), as well as 'the normal expected friendly neutral and allied shipping in US ports', the JCS could say that there was sufficient shipping to move US forces to Europe according to the existing schedule. This would be so, however, 'only if these movements are given priority over other US military requirements and over civil requirements during the first 60 days'.[6]

The assumptions and factors upon which this assessment was based included mobilization and commencement of hostilities beginning on the same day, prior to a pre-allocation of NATO shipping; the US national shipping authority will give preference and priority to US military shipping requirements from the pool of US ocean-going merchant shipping and 'allied and neutral shipping will be available for military use from the United States to Europe as follows':

Month	Passenger Space	L/T-Cargo
D+1	30,000	100,000
D+2	50,000	125,000
D+3	50,000	150,000[7]

It was estimated that 10 Army divisions would be deployed to Europe in the first 180 days: 2 in the first 30 days, 2 more in the next month, 3 in the following 30 days and 3 more divisions by D+180. Two marine divisions would be loaded onto ships of the Navy's amphibious fleet for an amphibious assault by D+30, with another marine landing at D+90. All US forces deployed, including Air Force and Navy, would require maintenance by sealift based upon a factor of one metric ton per man per month. It was anticipated that petroleum, oil and lubricants (POL) for the first 90 days of a war would 'be largely in place at the beginning of a war'.[8]

The capability of shipping from the US inventory allowed for the following factors:

a. *Transports*
   (1) Steaming time – 10½ days, one way.
   (2) Turnaround time – 46 days.
   (3) Sailing date of first convoy – D+13.
   (4) Convoy interval – 16 days for fast convoy.
   (5) Loss factor – 1¾ per cent of ship availability per month.

b. *Cargo Vessels*
   (1) Steaming time – 31 days (16 outward).
   (2) Turnaround time – 70 days.
   (3) Sailing date of first convoy – D+8.
   (4) Convoy interval – 8 days.
   (5) Loss factor – 3½ per cent of ship availability per month.[9]

As to the availability of shipping within the NATO framework, the JCS noted that 'no definitive answer can be determined concerning the availability of, or use of, shipping by the United States ... until DSA is activated and an initial allocation of NATO shipping, including the United States shipping is made by that body'.[10] Thus, while the JCS did assume that allied shipping in US ports would be available to the United States for military purposes, it was not at this time assuming availability of shipping through DSA in its calculations.

What the US military wanted was not just assurance that DSA would allocate ships in the event of war, but a 'pre-allocation' of NATO flag ships in order to allow the JCS to plan for sealift support of NATO's medium-term defense plan. As an interim step, the JCS requested that the Chief of Naval Operations (CNO):

> secure the necessary pre-allocation of US shipping to meet the urgent US military requirements for the early deployment of US forces for the first 90 days of a war ... This action, taking the indeterminate positioning of US ships into account, would provide the US planning to proceed, pending a pre-allocation of NATO shipping.[11]

It seems evident that, at this time, the US military was in fact skeptical of any NATO body being able to supply non-US flag shipping, and wary of the United States becoming dependent upon DSA. The JCS's Joint Military Transportation Committee, therefore, objected to including American military merchant transports as part of the US contribution to the NATO pool. A JMTC memorandum of March 1951 said that PBOS had been 'unresponsive' to US military requests and that it was trying to 'eliminate all military voice or representation in the organization' and had become too civilian. It was suggested that 'upon activation of DSA, or whenever a need for allocation of shipping ... became imminent, steps should be taken *at that time* for military representation within the DSA and to establish machinery so as to make operations of the DSA more responsive to the military organization responsible for the overall conduct of the war'. According to the JMTC the 'climate' was not right to 'achieve these military objectives'.[12] In the meantime, the JMTC would proceed with its own planning, based mainly upon available US government-owned (that is, the National Defense Reserve Fleet [NDFR]) and private US flag ships. It had already 'for the purposes of preparing transportation plans in support of war ... established adequate procedures for the control and processing of requirements and capabilities for ocean shipping and for making pre-allocations of shipping for the initial stage of a war'.[13]

As is described below, reservations about the adequacy of allied arrangements to supply vessels for a US sealift, and indeed doubts as to the reliability of individual national commitments, were frequently expressed by American military leaders (particularly when appearing before

Congressional committees on the US merchant marine). Since the United States had to consider extra-European conflicts wherein sealift would be required, it went forward with its own planning. Under the 1936 *Merchant Marine Act* subsidies were given to the US shipping industry in order to ensure adequate capability in time of war. The United States also had over 1,767 reserve ships, built during the Second World War, in the NDRF as of 1951 and over 2,000 in 1955.[14] Nevertheless, because the sealift problem was one of acquiring sufficient vessels quickly, and NDRF ships had to be broken out, there was an implicit assumption that the United States would be able, if needed, to draw upon NATO flag shipping plying the Atlantic in the event of an emergency. Whether this shipping would be allocated through DSA remained questionable.

Despite doubts about how DSA would function under mobilization conditions, PBOS commenced its yearly meetings and began to keep records of potentially available allied shipping. In 1952, the Alliance issued its first *Allied Control of Shipping Manual*. The manual, which was made available to all NATO flag shipping, offered detailed instructions for the conduct of merchant ships at sea in the event of war, including convoy procedures. According to Admiral Denny, NATO was 'firmly wedded to the convoy system and would seek to bring it about at the earliest possible moment' (Denny 1965: 362). To facilitate the turnaround time of convoys, Ship Destination Rooms were established in each Western European country in order to decide the destination ports for ships 'before the sailing of convoys, and the diversion of ships in convoy to other ports, if the original destination port became unable to receive them' (NATO 1955: 146). In addition, PBOS began to carry out a survey of the capability of European 'coastal and short sea shipping' (ibid.).

Thus, by 1955, it would appear that the Alliance had set in motion, under civilian direction, machinery to increase the collective readiness of NATO for sealift. However, since the single most important aspect of the allied sealift problem would be the rapid reinforcement and resupply of American forces, the relevance of PBOS planning, and indeed the relevance of non-US flag shipping, would depend on American assessments of their requirements and capabilities.

US SEALIFT ASSESSMENTS: 1954–74

Not only did the US military doubt the effectiveness of the allied civilian agency charged with allocation of sealift, PBOS, it was also concerned with the ability of the Maritime Administration (MARAD) to meet military requirements. In 1954, Secretary of Defense Charles Wilson and

Secretary of Commerce Sinclair Weeks concluded an agreement which had the effect of asserting greater military control over MARAD in the acquisition and control of ships for military purposes (US Dept of Defense 1954). The 'Wilson–Weeks Agreement' dealt with the 'utilization, transfer and allocation of merchant ships' in the event of a national emergency. A 'nucleus fleet' of over 200 ships was to be placed under the 'exclusive custody, jurisdiction and control' of DOD's Military Sea Transportation Service (MSTS). The agreement also outlined a procedure for augmenting this fleet short of full mobilization which gave preference to the US merchant marine. Under full mobilization, MARAD would provide, through the National Shipping Authority, additional vessels to meet DOD requirements as set forth by the Chief of Naval Operations. These additional ships would be drawn from the NDRF and from the requisitioning of private bottoms in accordance with the *Merchant Marine Act of 1936*.

The US military wanted to be assured of adequate sealift for the whole range of possible conflict scenarios. 'Faced with the unanswerable questions as to where another war might come, when, on what scale, involving what weapons and what allies, responsible officials ... tended to favor whatever seemed economically justifiable' (Lawrence 1966: 106). What this meant in terms of DOD, MARAD and shipping industry relations was that the military generally supported the Maritime Administration's requests for greater subsidies for the US merchant marine on national security grounds. MARAD's interest was in increasing support for the US flag fleet. The military, anxious to secure sufficient capability and aware that shipping subsidies did not come out of its budget, generally went along with MARAD's request (Jantscher 1975: 134). In making its case in support of greater aid to the US flag fleet, the military tended on the one hand to point out the continuing relevance of shipping to future conflicts, and on the other hand seemingly and deliberately to downplay the contribution which could be made by NATO flag shipping.

A 1959 study by the National Research Council Panel on the Wartime Use of the US Merchant Marine, 'Project Walrus', *The Role of the U.S. Merchant Marine in National Security* (US NRC 1959), supported military opinion on the potential usefulness of NATO's DSA. Since this NATO body was not expected to become operational 'until 60 to 90 days after the outbreak of a war involving NATO' (ibid.: 30), the US military could not rely on it for its early sealift. Therefore, the condition of the merchant marine had to be improved. However, the Panel was of the view that any future use of large-scale US military sealift would be for limited war purposes only – nothing larger than the Korean War. A Second World War situation was placed 'outside the range of possibility'. In its estimate of US requirements, therefore, the study did not consider the effects of attrition because a war in which shipping would be subjected to attack could no longer be considered limited. 'It is assumed,' the study pointed out, 'that the Soviets both know and understand that the indispensable condition of limited war is an inviolate base structure plus inviolate lines of communication' (ibid.: 4, 7).

Thus, in effect, the study supported the military's contention that merchant shipping was of continuing relevance, but cast doubt upon its role in a general European war. Indeed, a year earlier (1958) a DOD position paper on merchant shipping needs was rejected by the National Security Council (NSC) on the grounds that it failed to deal adequately with the impact of nuclear weapons on shipping needs (Lawrence 1966: 106).

The Department of Defense and MARAD found a more favorable audience in the House Armed Services Committee. In 1959, the Committee held hearings on the 'adequacy of transportation systems in support of the national defense effort in the event of mobilization' (US Congress 1959).

At the hearings, the office of the Deputy Chief of Naval Operations (DCNO) for logistics and MARAD produced a joint study which pointed to serious deficiencies in US shipping. With regard to the initial sealift of troops to Europe, the Navy acknowledged that its planning indicated that sufficient capacity existed in the NDRF and private active ships to meet the 'minimum requirements for troop transport in a general war'. Active and reserve inventories amounted to 156 ships, ranging in age from 5 to 30 years. Yet, the age and speed of most of these vessels made them inadequate for modern sealift or too vulnerable to submarines. Fully loaded, the 156 ships could move over 350,000 troops but only 12 of these (enough for 33,000 troops) could reach speeds exceeding 20 knots. And, of these 12, the 'fully modern lift potential' amounted to only 6, 'which added up to a total emergency capacity for about 14,000 troops, enough for one streamlined division'. There were 2 ships under construction, which would bring the modern capacity to 19,000 (ibid.: 49).

In a letter attached to the report, the DCNO Logistics, Vice Admiral Wilson, addressed the question 'as to whether our calculations of shipping needs for general war have taken into consideration the capabilities of our allies and the magnitude of multilateral commitments to the NATO pooling concept' (ibid.). His answer was 'a qualified yes'. The United States would abide by its shipping commitments to NATO in time of war by contributing shipping to the NATO pool because the needs of the allies 'would exceed their combined shipping capabilities'. As to whether the allies would be able, through DSA, to aid in the American reinforcement sealift at the beginning of a mobilization, Wilson was doubtful. European ships, at that time, were of marginal value under 'heavy load conditions'. Moreover, the DSA was not expected to become fully operational until two or three months after the outbreak of war. 'Meanwhile,' he noted, 'our immediate need for efficient transports must be filled from US flag shipping' (ibid.).

The US military wanted to assure itself of adequate shipping for emergencies other than a NATO war. For this reason, it would appear that the JCS was reluctant to place too much public confidence on shipping assistance from allies, lest critics of shipping programs use the possibility of allied assistance as a reason to cut back on these programs. It also seems that, even in terms of an emergency in Europe, the American military wanted to be assured of sufficient sealift in the event that a number of the allies, for whatever reason, delayed in shipping mobilization. The only reinforcement plan for NATO was the US plan. 'We must also recognize,' warned Wilson, 'that under various circumstances which might arise, the policies and views of the United States would not necessarily coincide with those of our many allies in each and every crisis. The lessons of history have taught us that, as a matter of fundamental policy, our primary reliance must continue to be placed on capabilities under our direct control' (ibid.).

Which capabilities could be counted as being under direct US control was a matter of some controversy, particularly in continuing efforts to subsidize the US flag merchant fleet under the 1936 Act. The government-owned MSTS and NDRF fleets would be available, although the latter would have to be broken out. Under the 1936 Act, ships from the private US flag fleets would also be available, in some cases before the bulk of the NDRF reserves. More questionable were those ships under effective US control, the EUSC fleet. These were ships owned by Americans, or US companies, but which were registered under flags of convenience, mostly Panamanian, Liberian and Honduran, the PANLIBHON fleet. As will be noted below, the US government did take steps to assure the availability of these ships in an emergency. However, the military and those lobbying for increased subsidies to the US shipping industry tended to discount the PANLIBHON fleet's availability just as it doubted NATO flag shipping availability. Even greater doubts were expressed about the availability of NATO-owned ships flying flags of convenience.

Military doubts about the NDRF were supported by a 1959 Government Accounting Office study, which found that the majority of the ships in the over-1,000 ship fleet were unprepared for rapid mobilization.[15]

MARAD did not even know the condition of most of the ships. 'Yet,' the GAO study argued, 'rapid activation of reserve vessels, which according to the joint MARAD–Navy Planning Group is the primary justification for their retention and preservation, would require that Maritime Administration have knowledge of the condition of each vessel' (US GAO 1959: 4).

This appears to have been one reason why the Navy supported so strongly efforts to increase subsidies to the US merchant marine. These ships would already be at sea, mostly serving US ports, and thus more readily available in an emergency. Lawrence notes that during the late 1950s and early 1960s, the CNO regularly supplied 'off the record' statements to the Chairman of the House Merchant Marine Committee which told of a 'marginal capability to carry out the requirements for general war' (Lawrence 1966: 107).

With the coming to office of the Kennedy administration and, more particularly, with the assumption of the office of the Secretary of Defense by Robert McNamara, a 'wholly new attitude' towards the problem of military sealift emerged (ibid.). This was especially the case with regard to European sealift requirements. During the 1962 hearings before the House Merchant Marine Committee, McNamara noted that, in the event of a European war, the United States would, despite enhanced airlift capabilities and prepositioning of equipment, employ sealift. He confirmed the military's dependence for early sealift upon the US merchant marine as a supplement to the government-owned fleets. He also mentioned the need for more military-suited ships such as roll-on/roll-off vessels (RO/RO).

McNamara, however, stressed that he did not wish to overstate the requirements, 'thereby providing an umbrella under which a huge ship construction program for the Merchant Marine can be justified'. He was 'reluctant' to recommend any program for further subsidization of the US commercial fleet 'other than one that is based solely on military requirements'. From a 'purely military point of view' he found that 'the reserve fleet plus the vessels in service, plus the construction program ... appear adequate to our needs' (US Congress 1962: 91–2).

McNamara did not specifically mention the availability of NATO flag shipping, but further administration testimony did point to NATO contributions to a US sealift effort. In his statement to the President's Maritime Advisory Committee (1965) the Secretary of Commerce supported McNamara's views, noting that 'existing ships under American flag are adequate to meet presently established requirements for services by the military'. He included in his assessment the NDRF and the EUSC fleets, the latter numbering about 450. Finally, he stressed that 'ships under the NATO alliance must also be taken into account' (US Dept of Defense 1965: 3).

A year later (1966) Maritime Administrator, Nicolas Johnson, told the House Merchant Marine Committee that DOD's sealift planning was based upon four assumptions: access to privately-owned US flag ships; access to ships in the reserve fleet; access to the EUSC fleets; and, 'under emergency conditions, when our established alliances are invoked, the foreign flag ships of the alliance will be available for allied tasks, including the needs of the Defense Department'. Foreign flags would also carry 70 per cent of US trade. Mr Johnson stressed to the Committee that the US 'has not equated our shipping needs with the active American flag Merchant Marine. Defense plans augment our active fleet with the reserve fleet, flags of convenience, our allies' ships and relies on foreign flag vessels' availablility for commercial trade – all at a cost to the government far less than subsidizing an active fleet' (US Congress 1966: 351–2).

Part of the new attitude towards sealift requirements, in addition to a more explicit inclusion of NATO resources, was the result of the enhanced importance of airlift and prepositioning of equipment. Under McNamara over $700 million was budgeted to produce 150 to 200 troop air transports per year, thus quadrupling the 1961 capacity by 1969. This would make the US air transport fleet equivalent to 250 to 350 Victory ships (Lawrence 1966: 110). Following the Berlin Crisis, enough equipment for two divisions and supporting units was prepositioned in Europe (US Congress 1962: 88).

In October 1963, the United States and NATO conducted exercise BIG LIFT, which involved the airlifting of 15,358 troops of an armored division from Fort Hood to the central front. The troops joined up with prepositional equipment, including over 12,000 tracked and wheeled vehicles.

The growing importance of airlift and prepositioning did relieve some of the pressure on immediate sealift in the event of a European war. However, because airlift could not accommodate the mass of equipment which would be necessary to sustain conventional resistance, concern with sealift continued and indeed, as is argued below, increased with the adoption of a more flexible response strategy by the Alliance. It is important to note, as well, that during a crisis period in which NATO was afforded some time between mobilization and commencement of hostilities, sealift was expected to play a greater role (in terms of tonnage moved) than airlift. A further point is that during the 1960s the bulk of the prepositioning was for the central front. On the northern and southern flanks, reinforcement would come from amphibious landing, necessitating sealift support.

The role of sealift in amphibious support of NATO operations was demonstrated in exercise STEEL SPIKE held in 1964 (US Congress 1965: 6). This USN/Marine exercise was apparently meant as an answer to BIG LIFT (and was also used by supporters of shipping subsidies before House Merchant Marine Committee hearings on the exercise) (ibid.). Nevertheless, the exercise did give an indication of the continuing relevance of sealift to possible NATO reinforcement operations. It was the largest peacetime amphibious exercise ever conducted in the Atlantic and the largest military landing 'since the Korean War'. Involved were some 28,000 Marines and their equipment, including tanks and trucks, along with 90,000 metric tons of supporting equipment. Much of the supporting equipment was moved from five US ports in merchant ships, some of which were government-owned and some requisitioned from private fleets, in a convoy that accompanied the amphibious force in a ten-day transatlantic crossing (ibid.).

In 1966, Secretary McNamara requested a comprehensive study of US sealift requirements and capabilities – *The Sealift Requirements Study* (US Dept of Defense 1966 and 1967). The first phase of the study was an analysis of the 'effectiveness and cost to government of alternative sealift postures to meet US national security requirements' in the 1970s and 1980s. These requirements involved regular peacetime DOD shipping and the requirements for a 'prolonged limited emergency'. In the second phase, the study considered:

> sealift requirements to support a major non-nuclear conflict in the NATO area ... it is to be assumed that requisitioning or seizure of the United States flag merchant shipping for support of DOD sealift would be authorized and NATO flag merchant shipping is available to support the total sealift effort.
>
> (US Dept of Defense 1966: I–1)

That portion of the analysis most germane to this study of NATO shipping was an examination of the requirements and capabilities to meet a European build-up, coincident with normal shipping requirements and conflict scenarios in Asia. It was assumed that the initial European build-up would take place, short of the outbreak of war involving the NATO countries, over a four-month period and that shipping would not be subjected to hostile enemy action (US Dept of Defense 1967: i, II–14). Only dry cargo shipments using break-bulk and container ships were considered in this portion of the analysis. The study focused on the capabilities in the late 1970s.

Table 5.3 presents sealift requirements in the event of a European build-up imposed upon an existing Vietnam- and Korean-scale conflict, again with the European build-up taking place short of the outbreak of hostilities. The study considered two alternative rates of build-up over a four-month period (annual rates of 12 or 19.5 million M/T) and two follow-on rates (annual rates of 9.6 or 16 million M/T). (M/T is a measured ton. Conversion from long tons, L/T: 2.71 M/T to 1 L/T for general dry cargo; 1.01 M/T to 1 L/T for bulk cargo.) In addition, there are two alternative sealift requirement levels for the support of NATO forces other than US.

Thus, to examine the scenario of a European build-up in 1966, given existing base military shipping requirements and the demands of Vietnam, a four-month build-up of supplies for US forces only, at a 'high' rate, would be 6.5 million M/T, with a 'low' rate of 4 million M/T. If the United States wanted to resupply its allies at the same time, the high rate for both would be 6.5 and 10 million M/T.

These calculations were based upon the assumption that the initial deployment and follow-on resupply of a 32,000 man division in 1967 would require 966,400 M/T at roughly 1.85 M/T per man per month. Projections indicated an increase by 1980 to 2.15 M/T per man per month (ibid.: II–9–II–10).

The study concluded that 'the most critical requirement for military sealift' by 1977 would be the situation in which an annual requirement of 28 million M/T (Korean War level) had superimposed upon it a sole US commitment in the NATO area. This would require an initial four-month lift of 6.5 million M/T followed by a 16 million M/T annual requirement. The build-up would correspond to an annual lift of 47.5 million M/T, and the follow-on to 44 million M/T. By comparison, the fiscal year 1966 Vietnam requirement was 20 million M/T. This US-only European resupply, plus a Korean situation, would be the most difficult 'which DOD might be called upon to satisfy without recourse to NATO shipping' (ibid.: i, II–27). It is assumed that in an American resupply effort which included equipment for allied armies, NATO flag shipping would be available.

The study concluded that the US flag fleet would become 'progressively less able to meet growing needs' and 'incapable of supporting large scale contingencies and the US economy at the same time' (ibid.: II–11–II–12). A building program of 25 to 40 dry cargo ships per year would need to be undertaken by the late 1970s in order to 'satisfy the largest postulated requirements without reliance on NATO shipping' (ibid.: iii).

Table 5.3 Representative sealift requirements*

Situation	Steady state (Annual)	European build up		Annual equiv.	Annual follow-on	Comparative total equivalent	
		Total	/months			Annual	Monthly
1 Vietnam (FY/66)	20					20	1.67
2 Korea	28					28	2.33
3 Korea + Vietnam	37					37	3.08
4 Vietnam +	20	4/4		12		32	2.67
NATO (US only)	20				9.6	29.6	2.47
5 Vietnam +	20	6.5/4		19.5		39.5	3.29
NATO (US only)	20				15.6	35.6	2.99
6 Korea +	28	4/4		12		40.0	3.33
NATO (US only)	28				9.6	37.6	3.01
7 Korea +	28	6.5/4		19.5		47.5	3.95
NATO (US only)	28				16.0	44.0	3.67
8 Vietnam +	20	4.0+		34.5		54.5	4.35
NATO	20	7.5	/4		18	38.0	3.17
9 Vietnam +	20	4.0+		42		62.0	5.17
NATO	20	10.0	/4		24	44.0	3.67
10 Vietnam +	20	6.5+		42		62.0	5.17
NATO	20	7.5	/4		18	38.0	3.17
11 Vietnam +	20	6.5+		49.5		69.5	5.8
NATO	20	10.0	/4		24	44.0	3.67
12 Korea +	28	4.0+		34.5		62.5	5.21
NATO	28	7.5	/4		18	46.0	3.83
13 Korea +	28	4.0+		42		70.0	5.83
NATO	28	10.0	/4		24	52.0	4.33
14 Korea +	28	6.5+		42		70.0	5.83
NATO	28	7.5	/4		18	46.0	3.83
15 Korea +	28	6.5+		49.5		77.5	6.46
NATO	28	10.0	/4		24	52.0	4.33

* Millions of measurement tons

Source: US Dept of Defense 1967.

The *Sealift Requirements Study* was used extensively in the legislative process which led up to the *Merchant Marine Act of 1970*, which was the most comprehensive overhaul to date of the 1936 Act. It was the intention of the new Act to provide a program of loans and subsidies to every sector of the US shipping industry in order to expand, modernize and increase the efficiency of the American merchant marine. The new effort moved away from general break-bulk cargo liners toward bulk carriers for oil, ore and liquefied natural gas. This construction would enhance the ability of the United States to meet its energy needs by utilizing American ships (Mayer and Handerson 1974: 6).

Reinforcement sealift, however, still relied upon break-bulk carriers for the initial deployment and container ships for the follow-on resupply. Break-bulks were especially necessary in the early stages. This type of ship was viewed as having several advantages 'considered vital in DOD contingency planning'. First, it has its own loading and unloading gear, necessary in the event that port facilities are put out of action or are unavailable. Second, it can accept most types of 'outsize' military cargo, such as the M-60 main battle tank, self-propelled guns and large trucks (which cannot be containerized). Third, break-bulk is well-suited to carry ammunition which containers have not been approved to carry (US Comptroller General 1976: 3).

In general, then, the program begun by the *Merchant Marine Act of 1970* was not expected to fill the growing gap between US sealift capabilities and the requirements for a European reinforcement. In 1972, the office of the Assistant Secretary of Defense for Program and Analysis released a major study which looked at American requirements and capabilities to meet a NATO sealift. The *Sealift Procurement and National Security* (SPANS) study (US Dept of Defense 1972) provided a 'definitive' analysis of sealift procurement needs, assuming a scenario of a conventional war in Europe followed by an Asian contingency. It identified sealift capacity required to meet military needs and compared these with the ability of the US merchant fleet to meet them.

The SPANS study provided the most optimistic assessment of NATO flag shipping availability to date. It concluded that roughly 200 European ships would be needed to supplement the US private and government-owned fleets as well as the effective US-controlled fleets (EUSC) (Mayer and Handerson 1974: 12). The study did not say that NATO flag shipping would automatically be available, but it did indicate that, given the amount of transatlantic shipping daily, and the kinds of ships involved (i.e. break-bulks), a significant pool would be at hand. Moreover, these ships, if committed to an American sealift effort pursuant to a mobilization, could be available in significant numbers before the NDRF could be broken out of reserve.

The study considered four different scenarios for the US merchant marine depending on progress in building programs and market penetration of the global commercial shipping industry. The most realistic scenario was the 'pessimistic fleet'. Numbering some 309 ships by 1976, this fleet would be the then current US merchant fleet, plus construction obligations then in existence (US Dept of Defense 1972: Pt II, Sec. B: 10–24).

The SPANS study also estimated the marshaling schedule of the 'pessimistic fleet' over the forty-five day mobilization period. Under 'late availability' conditions it was assumed that ships more than half way to their destination on M-day would proceed to those destinations before returning to military on-loading ports of the continental United States. It was estimated that approximately 122 US merchant marine vessels could be at the eastern and Gulf coast ports within days of a mobilization order. These would be the main ports from which reinforcements for Europe would originate. The majority of the ships available at M+10 would be break-bulk suitable for unit deployment, as well as about 50 RO/ROs. By M+15, an additional 25 ships could be at eastern and Gulf ports.

The augmentation of the US flag fleet by NATO ships during a mobilization would depend upon a collective decision by the North Atlantic Council and/or national decisions by member governments. The SPANS study assumed that the European nations would make their ships available as they became more concerned about Soviet intentions, with the need to demonstrate the Alliance's determination and to protect itself in the likelihood of war. Thus SPANS assumed increasing levels of NATO flag augmentation with increasing estimates as to the severity of the situation. At first, concerned but 'cautious' Europeans would make an additional 328 ships available. As they became 'worried' the number might rise to 442 (US Dept of Defense 1972: Part III, Vol. III: F–38–F–39). (Under NATO 'cautious' the total available shipping would be 637, under NATO 'worried' a total of 753 ships would be available to the United States.) Finally, 'scared' European governments might augment US sealift by 730 ships, bringing the total number of ships available to 1,039.

Utilizing a marshalling schedule under late availability (the same as that used for the US pessimistic fleet), SPANS estimated that under cautious conditions, nearly 100 more ships would be available to the US by M+10, an additional 40 under worried and nearly 150 more under NATO scared conditions (ibid.: F–47). In this case, significantly more

NATO flag shipping becomes available after M+30. This is a reflection of the greater time it would take the European flag ships to unload their cargoes and proceed to US ports. It is important insofar as early reinforcement depends on the availability of large numbers of break-bulk, but according to SPANS the majority of European break-bulks were not estimated to be available until a month into the mobilization.

If the SPANS study was overall optimistic about NATO flag shipping availability, it was also more favorably disposed to the inclusion of the EUSC fleet than previous DOD reports had been. It was noted that under the *Merchant Marine Act* the Secretary of Commerce could request or purchase any ship owned by a citizen of the United States. This applied to US ships flying in the PANLIBHON fleet, though the US, under international law, could not seize these ships on the high seas. However, neither Panama, Honduras nor Liberia prohibit a pledge by US owners of vessels flying under their flags that their ships can be placed in the service of another nation. 'In fact,' noted SPANS, 'many PANLIBHON registered vessels are contractually committed to the service of the United States in time of emergency in return for war risk insurance and as a condition for transfer of the vessel to foreign registry.' The PANLIBHON countries do not need these ships for their economies, nor could they protect them, nor could they stop the US from seizing them in US waters. SPANS concluded that 'it is unlikely that there would be any significant PANLIBHON interference with United States requisition of such vessels' (ibid.: G–9–G–10).

In addition to the use of NATO flag shipping and the EUSC fleet to augment the American merchant marine, SPANS assumed that the NDRF fleet, then numbering about 130 ships, would also be used in a European reinforcement. However, the study concluded that 'the subtraction of the NDRF' would not affect its overall favorable assessment of US ability, with NATO shipping, to meet a European contingency, although under attrition conditions its non-use would increase calculated shortfalls (ibid. Executive Summary: 4).

In its 'Executive Summary' the SPANS study noted that it had utilized the strategic assumptions 'commonly employed by the Services and the Organization of the Joint Chiefs of Staff' in arriving at its assessment. Nevertheless, the military did have significant reservations. In particular, the Army, the Navy and the JCS were 'concerned that the SPANS study could be interpreted improperly for future policy decisions, which could reflect an overly optimistic estimate of available sealift capacity'. Amongst those assumptions which the military found overly optimistic was the 'automatic availability of NATO shipping' (ibid.).

As in the past, the US military preferred to rely exclusively upon American flag shipping. By the early 1970s several factors combined to make this reliance, in a NATO contingency, almost impossible. The US flag fleet continued to decline, especially in the numbers of break-bulk available. In addition, the average age of the NDRF fleet in 1972 was 27 years old and it would take several weeks to break-out most of the ships. Most importantly, the adoption by NATO of a flexible response strategy, and American plans to support this strategy, re-emphasized under new conditions the central sealift problem. Now, more than before, early availability of shipping in sufficient capacity became the *sine qua non* of an effective and credible NATO reinforcement sealift posture.

FLEXIBLE RESPONSE: ENHANCING REINFORCEMENT SHIPPING AVAILABILITY

Since the beginning of the Alliance, there was concern with the problem of sufficient and timely sealift in order to support an American reinforcement of Europe. The formal adoption in 1967 of a strategy of flexible response brought about a renewed emphasis on the sealift problem. While the exact nature of a NATO and Warsaw Pact conflict could never be accurately predicted at any time since 1949, there was a general expectation by the late 1960s that the Alliance would attempt to meet a Soviet conventional attack at the sub-nuclear level. It was important, therefore, that the Alliance improve reinforcement capabilities. 'The strategy of flexible response,' wrote one analyst of the US defense transportation system, 'places great emphasis on the ability of the United States to rapidly deploy combat forces, equipment and supplies to Europe to counter an attack by Warsaw Pact forces' (Daniel 1979: 1). And, as a Ministerial meeting of NATO's Defense Planning Committee stressed with regard to conventional reinforcement under flexible response:

> Reinforcement and augmentation forces should reach an area of potential conflict before an aggression takes place, or if warning time is very short, early enough to affect the initial course of hostilities. Special emphasis should be placed on pre-stocking, on timely provision of sea and airlift and on adequate reception facilities.
>
> (NATO 1982: 72)

Three considerations, which had implicitly been at work in previous efforts to ensure adequate reinforcement sealift, became more salient with the growing emphasis on a flexible response posture.

The first of these was that steps to marshal shipping for a reinforcement sealift, and the peacetime plans to facilitate the timely availability of adequate capacity, were not dependent upon any single expectation as to the length or character of a crisis or conflict. To be sure, the beneficial effect of a sealift would be most pronounced the longer the time between mobilization and war and the longer the war remained conventional. Yet there would be no way to determine the effect of sealift at the moment mobilization began. The purpose of getting the transatlantic shipping pipeline going was to ensure that, short of an immediate escalation to the nuclear level, NATO would be in a position to sustain conventional resistance for as long as possible once the fighting began. Shipping itself would not determine the nature and length of a conflict, but without it conventional options would be restricted. 'As far as shipping plans are concerned,' noted the US Assistant Secretary of Commerce for Maritime Affairs in 1979, 'the debate of long or short war is not material.... We must plan for support of a war that begins with deployment of major forces in the shortest possible time' (US Congress 1980a: 89).

A second consideration was the relationship between air- and sealift. Since the early 1960s, the US and NATO had sought to improve rapid reinforcement through the prepositioning of equipment and the airlifting of troops who would take up this equipment. In a situation where a crisis arose unexpectedly and hostilities shortly followed, prepositioning and airlift would be essential to any kind of initial conventional resistance, particularly in the event of a short-warning Soviet attack. By the early 1980s, the US had prepositioned equipment for 4 mechanized divisions, with plans for an additional division prepositioning. According to a 1979 Congressional Budget Office Study, the US (using the then existing capacity) could airlift troops for 5 divisions to Europe in just over 2 days, with an additional 4 to 5 days required to draw the equipment and move to assembly areas. It would take just over 4 days to bring in 55 tactical fighter squadrons from the US (US CBO 1979: 49).

Improvements in airlift and prepositioning reflected NATO's concern over the growing capability of the Soviets to launch a short-warning, massive attack. But improvements in rapid airlift did not mean that the Alliance had officially accepted the assumption that a war would necessarily begin with little strategic warning. Indeed, one of the assumptions of flexible response was that NATO would have some warning. Airlift and prepositioning provided a hedge against the short-warning attack. The more time available between mobilization and war the greater scope for sealift in the reinforcement effort. Most of the personnel would still move by air, but nearly all of the cargo would move by sea.

Sealift alone could not only bring in the amount of re-equipment required for sustained conventional resistance, and thus improve the conventional balance along the central front, it could do so in a timeframe that compared favorably with that of airlift in the absence of additional prepositioning. For example, without prepositioning, it would take nearly 70 days to move 5 mechanized divisions to Europe after initiation of mobilization (ibid.). Sealift reinforcements, with equipment, could begin arriving in European ports after 14 days (Mako 1983: 68). Moreover, without prepositioning, additional airlift capacity hardly improved delivery times because of the difficulty of organizing an airlift of equipment as compared to sealift. The most modern Military Sealift Command (MSC) ship, the converted container ships SL-7s, could deliver 22,700 long tons whereas the largest transport aircraft, the C-5s, could deliver barely 132 tons. With speeds of 33 knots an SL-7 can make this trip from east coast ports to Europe in four days, and eight ships of this type would be able to load or off-load in one day 'the majority of the unit equipment – tanks, artillery, wheeled vehicles, helicopters – needed for two Army heavy mechanized divisions' (Kaufman 1981: 173; Holloway 1983: 34; Vlahos 1984: 21).

For these reasons American plans for the reinforcement of Europe assumed a 'full mobilization' of both airlift and sealift capabilities (US Dept of Defense 1976: VIII–4). As the airlift commenced there would be mobilization of the MSC's fleet, the US merchant marine and the pre-committed NATO flag ships, all of which would be needed to support a major NATO contingency (ibid.: VIII–2, VIII–3). Under these circumstances, according to a DOD report to Congress, *Conventional Reinforcements for NATO*, 'sealift complements the early movement of military equipment by air and is the dominant mode of sustaining them in any lengthy contingency' (ibid.: VIII–2).

The third, and most important, consideration is that the transit of reinforcements to Europe was based on American and not allied plans. In its planning NATO did take into account reinforcement commitments of the US, but NATO itself had no plan on how the air- and sealift was to be accomplished in order to assure reinforcement and resupply.[16] The American plans sought initially to reinforce and resupply US forces in Europe, in all sectors, and in accordance with CINCEUR's planned Time Phased Force Deployment data as approved by the JCS. CINCEUR is NATO SACEUR, but in this instance he would be acting in his American capacity only and would assume responsibility for US reinforcements once they reached Europe. American planning also included the movement of material from prepositioned stocks in the United Kingdom (US Dept of Defense 1976: VIII–8). (The withdrawal of France from NATO's integrated commands and departure of US forces from France necessitated the movement of some prepositioned equipment to England.)

The United States would, in the course of a sustained sealift, also resupply allied forces. But the initial and overriding task would be the reinforcement of American forces according to US plans. It is this build-up which would demand the timely availability of adequate shipping capacity in the early stages of a mobilization or war. The availability of such capacity will very much control the rate of build-up (Case 1973: 25–6).

In his 1974 report to Congress, Secretary of Defense James Schlesinger pointed out that there existed a 'substantial need for sealift capacity to sustain and augment forces initially deployed by airlift'. However, he stressed that the DOD-controlled fleet would be insufficient 'to support even a minor contingency in a timely fashion some years hence'. Heavy reliance would have to be placed on the US merchant marine 'and in the case of a NATO conflict, on the commercial fleets of our NATO allies'.[17]

THE READY RESERVE FORCE AND THE SEALIFT READINESS PROGRAM

By the mid-1970s, US Defense Department contingency planning called for supplementary sealift (to the existing MSC fleet) to be available 'within the first 10 to 15 days of a commitment of American forces or material'. A mobilization order by the President would place the US merchant marine at the disposal of the MSC. However, the number of break-bulks in this fleet had declined significantly. According to a 1976 report by the US Comptroller General: 'Because of the shortage of break-bulk ships, the 130 Victory ships ... maintained by MARAD in NDRF represent a capability that is considered vital in DOD contingency planning' (US Comptroller 1976: 3).

But the problem with the NDRF was that it could no longer meet DOD sealift requirements in the timeframe said to be necessary. For example, at the Beaumont, Texas, site the estimated time required to activate a Victory ship ranged from 18 to 36 days (ibid.: ii). In order to enhance the early availability of US flag shipping, the United States initiated the Ready Reserve Force (RRF) and Sealift Readiness Program (SRP). In the RRF, the goal was to improve the 'immediate surge capacity' of the MSC by upgrading ships from the National Defense Reserve Fleet so that they could be 'at berth' within ten days of an activation of the NDRF. By 1980, 23 ships were available under this program (US Dept of Defense 1980: 18).

Creation of the RRF also allowed the MSC to buy ships from the

American flag merchant marine. By the early 1980s, US flag ships were carrying less than 5 per cent of the American seaborne trade and continuing to decline as an industry. Rather than maintain the fleet through higher subsidies, the MSC was able to acquire idle ships at relatively low prices and convert them for permanent use by the USN. This was the origin of the SL-7 fast sealift ships. These ships, which numbered eight by 1984, were formerly container vessels owned by Sea-Land services of New Jersey. They were converted into cargo carriers and vehicle carriers. Including these ships, the RRF numbered 31 by 1984 and rose to 77 by 1988 (Vlahos 1984: 21; *Jane's* 1983: 724–5). Part of the reason for this build-up was renewed emphasis on the need for rapid sealift to support operations outside the NATO area, particularly in the Persian Gulf. Several SL-7s might be on permanent station, fully loaded, ready to support deployment of the Rapid Deployment Force.

As a subgroup of the NDRF, the RRF could still not be fully activated until requisitioning authority had been obtained pursuant to a presidential declaration. However, MARAD and the MSC operate under the assumption that in NATO and non-NATO contingencies 'if the decision is made to commit a force ... the less difficult decision to activate the ships to deploy that force would likely follow' (US Congress 1980a: 15).

Since the RRF and most of the NDRF would have to be brought up to sea-going status, the SRP was also initiated to provide an even more rapid augmentation of ships under MSC control. Under this program commercial carriers would commit at least 50 per cent of their ships to requisition in time of emergency, in exchange for the right to bid on peacetime US military cargo. The plan called for the ships to be available in the first 60 days of a mobilization, with 20 per cent in the first 10 days, 30 per cent in the next 20 and the remainder in the following 30 days. By 1980, the MSC had obtained commitments on 200 dry cargo and 70 tankers (US Dept of Defense 1980: 18).

These improvements, however, would still have left the US short on sealift capacity for a European reinforcement within an acceptable timeframe. The American private and government-owned fleets did have sufficient capacity to deliver the necessary DOD tonnage in about 150 days, 'with the bulk of the necessary tonnage being delivered in less than 100 days'. Yet, as a MARAD administrator told a Congressional hearing in 1979, DOD required that deliveries be made in 'a much shorter time, a time too short to allow most ships to make repeated voyages. To meet the DOD timing requirements ... enough ships are needed to lift the load of unit equipment accompanying ammunition and resupply at one time.' This could not be accomplished by the US alone. The 'shortfall' would be made up by using 'high performance European cargo liners for movement of reinforcement shipping'. These ships would become available to the US when the North Atlantic Council requests reinforcements. 'The administrative organization which would co-ordinate the selection and chartering of the ships by the United States has been designed and exercised from time to time' (US Congress 1980a: 90). In his 1980 report on American mobilization capabilities, Under Secretary of Defense Robert Komer stressed that 'in a NATO emergency the largest potential augmentation of US sealift capabilities would come from other NATO countries' (US Dept of Defense 1980: 19).

THE NATO SHIPPING CONNECTION RE-EMPHASIZED

The increased importance of the timely acquisition of shipping required to support a flexible response, combined with the decline in American shipping capabilities, resulted in a concerted re-emphasis upon NATO shipping availability in the late 1960s and early 1970s. In 1966, PBOS began a general review of its operations and plans. By 1971, the US Maritime Administration was reporting that major progress had been made in the 'revision of NATO military plans for Navy control and protection of shipping' (US Maritime Administration 1972: 50).

More significantly, the US Departments of Defense and Commerce took steps to ensure the kind of preallocation of NATO flag shipping which the Joint Chiefs had sought in the early 1950s. In September 1973, DOD and Commerce entered into a new agreement with NATO 'to increase the availability of NATO flag shipping in the event of a major deployment of forces to Europe'. Emphasizing again that the sealift problem was 'not so much a matter of total capacity as early availability', Secretary of Defense Schlesinger reported to Congress in 1974 that 300 'suitable' NATO flag ships (some RO/RO, but mostly break-bulk) 'which normally frequent US east and Gulf coast ports would be "earmarked" in peacetime to facilitate their acquisition in a contingency'. These ships would be directed to American ports 'in response to specific US deployment requirements'.[18]

The key element in this new arrangement was that it would give the US MSC access to NATO flag shipping during a period of rising tension or immediately upon the outbreak of war, 'prior to the establishment of the Defence Shipping Authority'. The identified ships were to be considered as being available to the United States when the North Atlantic Council requested reinforcements (US Dept of Defense 1981). By the late 1970s, the pool of specifically earmarked NATO flag ships had been increased to 600 dry cargo, mostly break-bulk vessels. As a representative of the JCS Current Plans Branch-Logistics Directorate told a hearing on Defense Sealift Capability:

> our European allies have agreed to augment the US flag initial lift capability with 400 European NATO cargo liners in a NATO mobilization situation. To insure timely availability of the requested 400 ships, over 600 have been earmarked as a specially controlled reinforcement pool. The Maritime Administration accomplished this for us through its participation in NATO's Planning Board for Ocean Shipping and MARAD maintains an up-to-date listing of these ships by name and type.
>
> (US Congress 1980a: 15)

In 1977, the major NATO commanders completed a series of reinforcement studies aimed at identifying measures for improving airlift and sealift. The studies were then considered by NATO's Senior Civil Emergency Planning Committee and its subordinate bodies, Planning Board for Ocean Shipping, Civil Aviation Planning and the Planning Board for European Inland Transportation. The sealift portions of the studies recommended that further steps be taken to insure that ships earmarked for reinforcement were made available early in a crisis and that these ships should include the most militarily useful vessels, such as roll-on/roll-off, barge carriers and large fast break-bulk ships (US Senate 1978: 1651).

In November 1977, PBOS reported on these recommendations, agreeing on the special earmarking of ships which would be available in a crisis and calling upon allied governments to legislate the authority to obtain such ships on short notice, if such authority did not already exist. The report was considered by SCEPC in December 1977, and was viewed as 'a major milestone in NATO efforts to establish improved capability to reinforce' (ibid.). Also undertaken at this time was a wholesale revision, and streamlining, of PBOS procedures.

Between 1977 and 1982, the Alliance undertook a total review of its reinforcement plans. In December 1982, the Ministerial meeting of the North Atlantic Council approved an Alliance-wide plan for the crisis reinforcement of Europe with forces from North America, Portugal and the UK. The plan called for the movement of hundreds of combat aircraft, hundreds of thousands of troops and millions of tons of equipment, ammunition and resupply to Europe within a period of several weeks. Most of the personnel would travel by air while nearly all of the cargo would be brought by sea.[19]

In order to facilitate American access to the earmarked NATO shipping before the establishment of the DSA (West), a new NATO shipping body had already been created in 1977 – the Civilian Sealift Group (CSG). The CSG was a group of representatives from the major NATO shipping companies who would assist the United States in marshaling the 600 NATO flag ships earmarked for a reinforcement sealift. It would maintain an updated (each April) list of these ships. In the event of a decision to commence reinforcement, the MSC would first utilize available American shipping. If, however, it exhausts US assets or it finds that ships on the NATO list are more readily available, the MSC would advise the CSG, through MARAD, that it requires additional assets for reinforcement of the NATO front, and request that the CSG provide them. The MSC would initially request those ships from the 600 list especially designated to be used first, roughly 300 ships. Once this group was utilized, it would return and request the CSG to supply the remaining ships on the list drawn from NATO nations on an equitable basis.[20]

In terms of levels of tension, the NATO civilian shipping organization could be viewed as shown in Figure 5.1. Figures 5.2 and 5.3 show the sequence of shipping mobilization which would be undertaken by the US and how it relates to the acquisition of NATO flag shipping.

The steps to bring US government-owned, US private and NATO flag shipping to a state of readiness and at the proper location to support an American reinforcement sealift would be taken simultaneously with the commencement of airlift and the activation of civilian airlift capacity. This would necessarily have to be the case since the airlifting of troops to

Level	Body	Function
Peace:	PBOS	Planning, maintains lists of all allied shipping
Tension:	CSG	Supply of early reinforcement shipping
War:	DSA	Organization of all shipping both for follow-on reinforcement and resupply and for support of civilian economies
	DSEB* (West)	DSC** (East (UK))
		Southern Europe Group

*Defence Shipping Council
**Defence Shipping Executive Board

Figure 5.1 NATO civilian shipping organization

prepositioned equipment could not be counted upon as being sufficient to sustain conventional resistance. Thus, the sealift 'pipeline' would have to be set in motion with the initial acquisition and direction to US ports of available capacity. Another controlling factor would be the level of tension and the political situation internal to the Alliance. At low levels of tension, and in advance of a call by the Atlantic Council for reinforcement, the United States would first activate and draw upon its own resources.

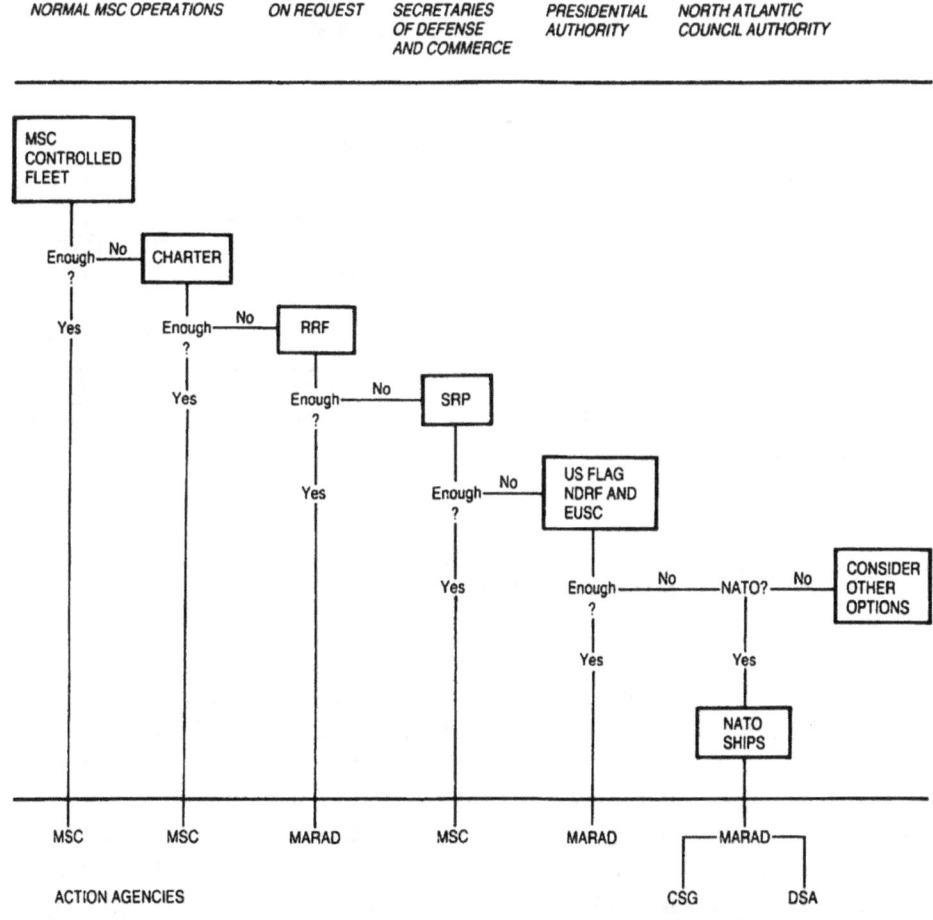

Figure 5.2 Sequence of acquisition of US shipping
Source: US Dept of Defense (1981).

154 Seapower in the nuclear age

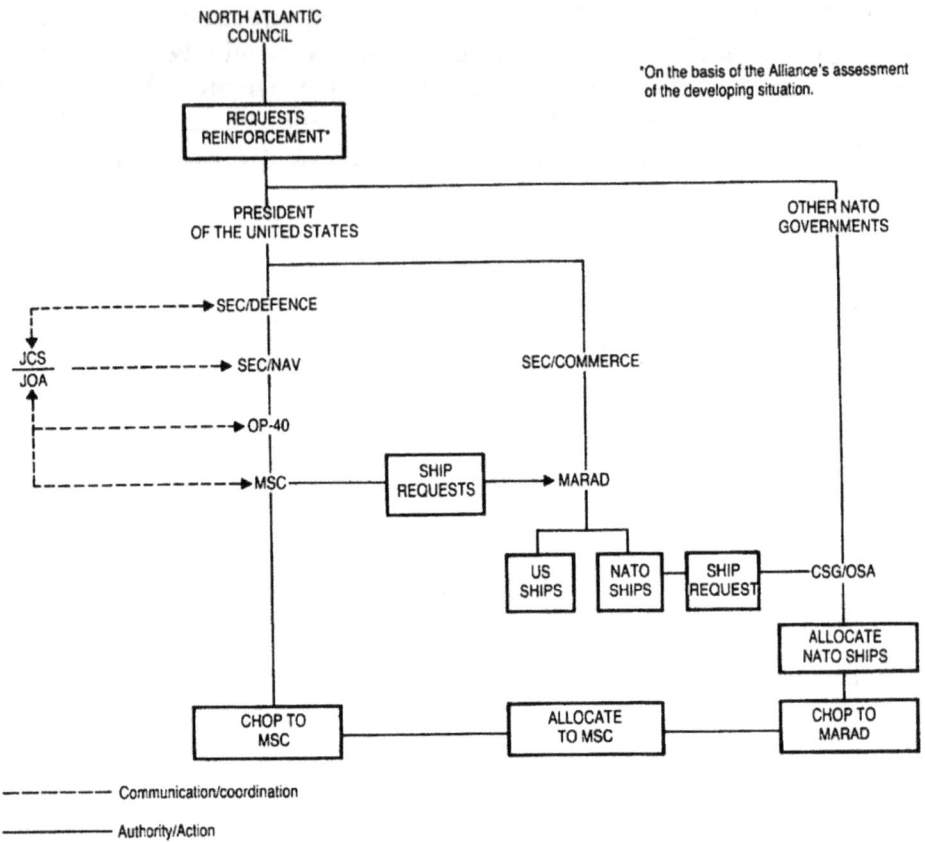

Figure 5.3 Acquisition of European NATO ships
Source: US Dept of Defense (1981).

Initially, this would mean using ships already under MSC control and the airlift capacity of the Military Airlift Command (MAC). The MSC would also draw upon US private shipping, voluntarily made available. At this point, the level of tension may be such that the US does not want to proceed with a full-scale mobilization. Here, it may wish to signal determination in a crisis, such as took place during the Berlin Crisis of 1962. Thus, rather than take ships out of commercial use through requisitioning, an attempt would be made to draw additional shipping from the RRF. As tension mounts and as the amount of time expected before the outbreak of hostilities decreases, aircraft committed to DOD under the

Civil Air Reserve Fleet (CRAF) program's Stage I would be requisitioned on the authority of Commander-in-Chief MAC. Designated SRP vessels would be next with a gradual call-up 'if possible to minimize industry impact'; at roughly the same time, CRAF Stage II would be put into effect on the authority of the Secretary of Defense. Finally, upon presidential declaration of a national emergency or mobilization, the final stage of CRAF would be implemented, the NDRF activated and the EUSC fleet ordered to return to the United States. Depending upon the amount of shipping which is available at this point, NATO flag shipping would then be requested through the CSG (US Congress 1980a: 17).

In the initial stages of a reinforcement effort, airlift capacity would be much more available, indeed hardly any sealift capacity would be moving. Thereafter, as noted, sealift would carry the bulk of the equipment for a European reinforcement. While estimates varied according to assumptions, particularly regarding the warning time and time between mobilization and the outbreak of hostilities, by the early 1980s the US Army expected to move 11 divisions to Europe between 30 and 90 days after mobilization through a combination of air- and sealift (Mako 1983: 53).

Requests for all ships from the NATO shipping pool would be made by the US Military Sealift Command through MARAD, according to methods outlined by NATO's Military Committee in its Procedures for Submission of Military Requirements for Ocean Shipping in Time of War, MC86/3. This document set forth a standard request form to be used by all National Shipping Authorities, especially MARAD. This form is reproduced in Figure 5.4 and followed by a sample of a completed one in Figure 5.5.

Ship type A included: roll-on/roll-off (RO/RO); break-bulk freighter (BB); lighter aboard ship, barge carrier takes on and discharges barges acting as a shuttle between ports (LASH); SEABEE, another class of barge carrier; refrigerated break-bulk; container (either self-sustaining, i.e. with its own unloading capability or non-self-sustaining); combination (container RO/RO or freighter RO/RO). Capacity (C) related to the total tonnage to be moved or the number and size of containers. Cargo types were as follows: ammunition; POL; containerized dry cargo; non-self-deployable aircraft; tracked and wheeled vehicles; refrigerated container cargo; refrigerated break-bulk; outsized cargo; and heavy lift.

In the case of a European reinforcement lift, the destination or port range would be one of the following: northern Europe, Bay of Biscay, Atlantic coast of Spain, Western Mediterranean, Eastern Mediterranean. The northern ports, particularly those of Holland and Germany, have the best port facilities. As of the mid-1970s, French ports in the Bay of Biscay were still included as possible off-loading sites of reinforcement. The expectation was that in a real war or crisis, the French might allow, and indeed may want, reinforcement and resupply to move through their territory.[21] The selection of a particular port range would be made on the

FROM	COMSC TO MARAD BY DIRECT LINE				
DTG					
CLASSIFICATION					
ALLOCATION REQUEST NUMBER					
DESTINATION (PORT RANGE)					
(TYPE) A	(NR SHIPS) B	(CAP*) C	(CARGO) D	(POE) E	(ON BERTH) F
1					
2					
3					
4					
REMARKS					
1					
2					
3					
4					

*BB   00s of MT
Ammo   00s of ST
POL   00s of Lt
RO/RO   000s of square feet
Cont   By size and quantity

Figure 5.4 Ship request work sheet
Source: US Dept of Defense (1981).

basis of the Operational Plan (OPLAN) which the reinforcement shipping was intended to support.

Where possible, the European members of the CSG would be allowed to select vessels from the country of destination (or one adjacent) for a particular lift (US Dept of Defense 1981: 21). However, the US MSC, acting as the sealift arm of the JSC in support of American reinforcement plans (or decisions made at the time of crisis), would want considerable discretion in the selection and use of ships from the NATO pool. Ships types had to meet MSC's immediate needs. For this reason, US MARAD would take the lead in identifying NATO earmarked ships in transit at the outset of a crisis, estimate location and arrival at destination, its cargo type and quantity, and forward the information to its MSC, which would in turn select or reject the acceptance of the vessel.

291201Z Jul 80

CONFIDENTIAL (CLASSIFICATION FOR ILLUSTRATION ONLY)

MSC (PROJECT CODE)-001

RANGE (A) OR (BELGIUM–NETHERLAND)

	A	B	C	D	E	F
1.	BB	1	110	VEH/GENNOS	MHC	30 JULY
2.	RR	1	80	TRACK VEH	WIL	30 JULY
3.	Cont	1	850	GENNOS	BAL	31 JULY

REMARKS

1. 70 TON CAPACITY AT ONE HATCH. 20 KNOTS.

2. SIDE LOADING RAMPS FOR XM-1 TANK. CARGO INCLUDES 14 TRAILER-MOUNTED 20 FOOT VANS OVERALL HEIGHT 11 FEET. ADVISE IF VANS CAN BE STOWED BELOW DECK.

3. SELF-SUSTAINING WITH SERVICE SPEED 22 KNOTS OR BETTER. DOD WILL SUPPLY 20 FOOT CONTAINERS AND DELIVER SHIPSIDE BY RAIL.

Figure 5.5 Ship request format (example)
Source: US Dept of Defense (1981).

Moreover, because ships especially suitable for military operations, such as RO/ROs, barge carriers (i.e. LASH and SEABEE) and high-speed, break-bulk carriers, would be limited, the MSC would want to collect them into a fleet rather than send them out on a ship-by-ship basis as they become available from the NATO pool. As DOD noted, the purpose of this would be to insure that 'these ships are used to best advantage during the course of the on-going operation'. For example, the US military would want to avoid the situation where all available RO/ROs were committed for 'opportune cargo before an armored division moves'. MARAD would furnish MSC with a 'projection of the availability of US and, as appropriate, European/NATO RO/RO and barge carriers, as soon as possible after the alert order is issued' (ibid.: 24).

The procedures outlined above were designed to provide for the transfer of NATO flag shipping in support of an initial American reinforcement effort in accordance with OPLANS, formulated by the JCS. This reinforcement effort would take place upon alert and request for reinforcement by the Atlantic Council during a period of tension. If the crisis continued without open hostilities, or if NATO was engaged in a protracted conventional war, on any front, then the procedures for sealift capacity acquisition would change.

As soon as is possible 'after execution of an OPLAN is initiated' the US Navy would begin informing MARAD of its monthly projections of future shipping requirements. Sustaining support shipping would not necessarily be operated under MSC control. Depending on the intensity of the crisis and the time factor, DOD might seek to move cargoes on vessels operated by the NATO Defence Shipping Authority, vessels drawn from the overall NATO pool (i.e. beyond those in the specially earmarked CSG fleet). Indeed the augmentation of the MSC fleet through transfers from the NATO pool was not to exceed '25 per cent of the total military requirements, as determined by the JCS'. The US Defense Department would reserve space on DSA-operated vessels for its cargoes which would normally be given priority. DOD would pay set rates for this usage.

In circumstances where DOD had a special operation during the period of sustained support, MARAD would, through the DSA, assemble European shipping to supplement shipping already under MSC control (ibid.).

For the initial augmentation of US capacity by NATO ships, and for the follow-on sustaining support, it was necessary to have as comprehensive a list as possible of shipping under control of the allied nations. Thus in the early 1970s the US military, through its own MARAD and through PBOS and SACLANT, began to press for better tracking of allied merchant ships in peacetime. In 1975, the United States initiated a PBOS general review of plans for wartime shipping operations, in order to refine operating procedures to promote the effective use of computers in tracking and locating available and useful shipping. This would include not only NATO flag, but PANLIBHON vessels owned by NATO nationals

In 1977, SACLANT sponsored a joint DOD/USN/MARAD/NATO study entitled *World Fleet Position Report*. Questionnaires were sent to all maritime nations in order to obtain specific data on ships which could be vital to American strategic logistics, economic or security interests. The final report gave the names of vessels, their country of registry, home port, cargo, frequent ports of origin and destination, speed and approximate longitude and latitude as of 1 June 1977. SACLANT noted that 'if circumstances demanded, a report of this nature could be generated in 48 hours. If only those ships deemed to have military capabilities were plotted, the lead time would be considerably shortened.' A list of ships in the EUSC fleet had already been maintained by SACLANT, with a 50 per cent 'overbid' to account for 'malpositions and lay-ups' at the moment of requisition. Computerized tracking of the entire world fleet by MARAD could give SACLANT daily reports by the mid-1970s (US Congress 1980a: 71–2).

By the late 1970s, the US and NATO were conducting joint 'command post' exercises to test procedures for making NATO shipping available to the US, particularly the earmarked ships. NATO conducted Shiploc-tests, which examined procedures for locating the earmarked ships, and in 1979 the NATO/JCS WINTEX/CIMEX exercise tested reinforcement and ship marshaling procedures. In conjunction with the latter exercise, the Civilian Sealift Group was exercised (ibid.: 108).

It is evident that, in general, the advent of flexible response did result in enhanced efforts by NATO to meet the critical problem of any American sealift reinforcement – the early availability of the right kind of shipping. In total, the allied nations owned over 10,000 ships of all types. It was being estimated that by the mid-1970s from 2,500 to 6,000 of these would be necessary for the initial deployment and follow-on, depending on the mixture available. The remaining ships would support the allied economies but at reduced levels of consumption.[22] Estimates of ship sinkings by Soviet submarines would push the number required higher. By the late 1970s, however, the US military could really count on a maximum of 600 NATO flag ships. There were no specific procedures beyond the DSA system for access to more specially earmarked allied ships.

Thus from the perspective of the US military, particularly the Navy and the MSC, the situation in the 1970s regarding reinforcement shipping remained unsatisfactory. Beyond the lack of any definite procedure for obtaining access to greater numbers of allied ships, several problems were being cited.

Almost 25 per cent of the NATO pool consisted of flags of convenience – PANLIBHON ships – whose crews were mostly non-NATO nationals. Unlike the EUSC fleet, these ships were not legally obligated under the *Merchant Marine Act of 1936* to be made available to the United States during an emergency. A further problem was that even those ships registered under NATO flags might be held back by the National Shipping Authorities in each country, in order to meet the needs and demands of the national economies. There was also no agreement on who pays for ships requisitioned from the NATO pool in the event of a mobilization which is subsequently called off. Ship owners would be reluctant to commit their ships under these circumstances.[23]

A much more general and far-reaching problem was the increasing containerization of all the NATO flag fleets. In terms of overall capacity, containerization compensated for the decline in the absolute numbers of ships. Between 1971 and 1980, the number of general cargo break-bulk ships sailing under NATO flags declined from 9,790 to 7,360. Tonnage, measured in dead weight tons (dwt), declined from nearly 48 million dwt, to just over 39 million dwt. However, during this same period the number of container ships and dwt increased from 182 ships totaling 2,314,000 dwt to 284 ships with a total dwt of 5,901,000. Of this latter ship total, the US had 87, Britain 74 and Greece only 7 container ships (Davy 1982: 68).

Containerization, with its larger tonnages for each ship and quicker loading and off-loading, would be of great value in a reinforcement effort. It would reduce the number of ships needed (in general, a container ship can replace a break-bulk, general cargo ship on a ratio of 7 to 1), solving the main problem of early availability of sufficient capacity, and cut down on turn-around time. However, under war conditions as opposed to a reinforcement carried out during a time of tension, containerization could well prove difficult, dangerous and costly.

In the first instance, container ships require elaborate unloading facilities, available only at select ports. These could easily be put out of action with even limited bombardment or sabotage.[24] There are no emergency port facilities for container ships to compensate for early loss of existing off-loading equipment. One of the objections which the military had to the conclusions of the SPANS study was that it placed too much reliance upon emergency off-loading equipment for container ships, which might not be available (US Dept of Defense, Executive Summary:

3). According to one US Army logistics expert, 'if emergency equipment is not selected and stockpiled before mobilization day, loss of the major European ports could mean loss of a European war' (Case 1972: 51).

Perhaps of greater danger and cost would be the effect of a Soviet anti-shipping campaign on a reinforcement and resupply effort based upon container ships. In a wartime environment 'a fleet consisting of a relatively few high productivity ships affords no cushion for combat losses'. During the Second World War, the sinking of a dozen Victory ships was not a mortal blow; however, one container ship with hundreds of tanks could 'harm' the entire allied effort (ibid.). Containerization, combined with advances in naval weaponry, could make it easier to inflict an unacceptable level of attrition on the American sealift reinforcement: 'A fleet of 18 large container ships can do the work of 120 conventional ships: only 18 missiles are needed to do the work of 120 torpedoes in the last war' (Davy 1982: 74).

By the late 1970s, increasing containerization was also undermining, to a certain extent, the utility of the arrangements that had been made earlier in the decade to place several hundred NATO flag ships at the disposal of the United States in advance of the activation of DSA. Quite simply, as the number of European break-bulks declined, the number of this type of ship available also declined.

NAVAL CONTROL OF SHIPPING AND CONVOYS

In order for merchant shipping to be placed at the disposal of a combined allied effort, to augment land forces during a time of tension or to sustain them in the event of hostilities, naval control of shipping must be established. Under emergency conditions, naval commanders direct all ship movements, in addition to providing protection for merchant ships, both civilian and those involved in a military sealift. It will be the naval commanders as well who will organize and escort convoys.

NATO and its member governments developed plans and procedures to establish naval control of shipping on an allied basis arid to integrate these procedures with allied and national civilian shipping authorities. The Supreme Allied Commander Atlantic (SACLANT) was designated by the Military Committee as the lead MNC (Major NATO Commander) for all naval control of shipping operations, including those in the Mediterranean. SACLANT established a NATO Shipping Working Group (NSWG) composed of naval officers from various NATO commands and from allied navies individually. Reflecting the growing concern with sealift that accompanied the adoption of flexible response policies, the

NSWG developed plans and recommendations to control shipping in the event of emergency or war. It maintains close liaison with PBOS, and representatives of both organizations attended each other's meetings.

Although the exact plans developed by the NSWG remain classified, it is possible to examine the basic steps that would take place in the establishment of NATO-wide naval control of shipping. As with the acquisition of sealift capacity, these steps would be tied to the existing level of tension that confronted the Alliance.

In the first instance, it is important to note that most NATO navies maintain Naval Control of Shipping Organizations (NCSORG), or at least the capability to bring one into being at the earliest possible moment. In the United States Navy the Commander, Military Sealift Command (MSC), is the primary officer responsible for naval control of shipping. Each of the Commanders-in-Chief, however, has established NCSORGs, and they operate under his command. CINCLANT maintains in peacetime a skeleton organization with NCSORG offices in American ports as well as all major European ports. These offices would be responsible for American flag shipping under less than full mobilization conditions. They also prepare for rapid expansion so as to be able to execute their tasks upon a declaration of mobilization by the Atlantic Council. At that point, the USN's NCSORGs would, in conjunction with European NCSORGs, direct allied shipping (Bender 1982: 7).

In the event of a sudden Soviet assault along the central front or on the flanks, coupled with the movement of Soviet anti-shipping forces onto the high seas, SACLANT might order the seas cleared of all allied shipping except for especially protected reinforcement convoys. The expectation, however, has been that hostilities would emerge out of a crisis situation. During a time of rising tensions, commanders, acting either in their NATO or national capacity, would declare certain sea areas, i.e. the Baltic, to be Merchant Shipping Control Zones (MERZONE). In these areas, ships already in transit would be guided, supervised and protected by allied naval commanders.[25]

At this time, depending on the degree of national and collective threat perceived, each nation would activate its civil direction of shipping, the National Shipping Authorities (NSA). In the US this would be the responsibility of MARAD. The NSAs would 'assume custody and control of all national flag merchant shipping' and would be responsible for 'all matters related to the employment and operation of merchant ships'. This would include: allocation, i.e. to NATO or national military authorities, itinerary, maintenance and manning. These operations would be similar to those performed by the ship owners in peacetime, but 'in time of war, the national effort is directed to fulfilling the Alliance's requirement for ocean transport in the prosecution of the war' (US Dept of Defense 1973: 1–1). These NSA tasks would be carried out under the direction of the NATO Defence Shipping Authority once this body was established. Each nation's NSA would designate representatives at key ports and establish Ship Destination Rooms at those ports. Figure 5.6 shows the civilian shipping organization for Germany.

The NSAs and the DSA are mainly concerned with the supply of shipping for military and civilian purposes. Control of shipping would be a naval responsibility under the command of a NATO area commander, who is also often a national area commander. In Figure 5.7, NCSORGs are shown as being under German naval commanders serving in allied capacities under COMBALTAP.

In each allied area, the national NCSORGs would be meshed together under the authority of a designated Operation Control Authority (OCA) who would be responsible 'for the control of movements and protection of allied merchant ships within a specified geographic limit'. This would entail convoy organization and routing, independent ship movements and diversion of shipping. In addition, the OCA would co-ordinate ship movements with area naval commands, maintain communications, intelligence and piloting systems 'to ensure rapid and secure dissemination of operational information'. All Naval Control of Shipping Offices (NCSOs) and Reporting Offices (REPTOFS),[26] which would control ship movements in port areas, would be under the OCA's command (ibid.).

The speed and sequence of activation of the national NCSORGs, and the assumption by allied OCAs of control over the national organizations in given geographic areas, would depend on the level of tension and threat perceptions. The most critical factor would be movements of the Soviet northern and Baltic fleets. If significant numbers of Russian submarines were detected moving toward the shipping lanes in the Atlantic or the Mediterranean, the pace of implementation of NATO-wide naval control of shipping would likely quicken.

It is possible that in the early stages of a crisis, and in the absence of significant Soviet naval movements, some allies would not activate their NSAs or NCSORGs. The United States could unilaterally place its NCSORGs on partial mobilization status as it began to marshal its own shipping in preparation for a reinforcement sealift. This would include CINCLANT's NCSOs at key European ports. Yet, it seems exceedingly unlikely that the US Navy would assume control of shipping without being joined by at least some allies, especially the British who are still a major participant in PBOS and very dependent upon shipping, also

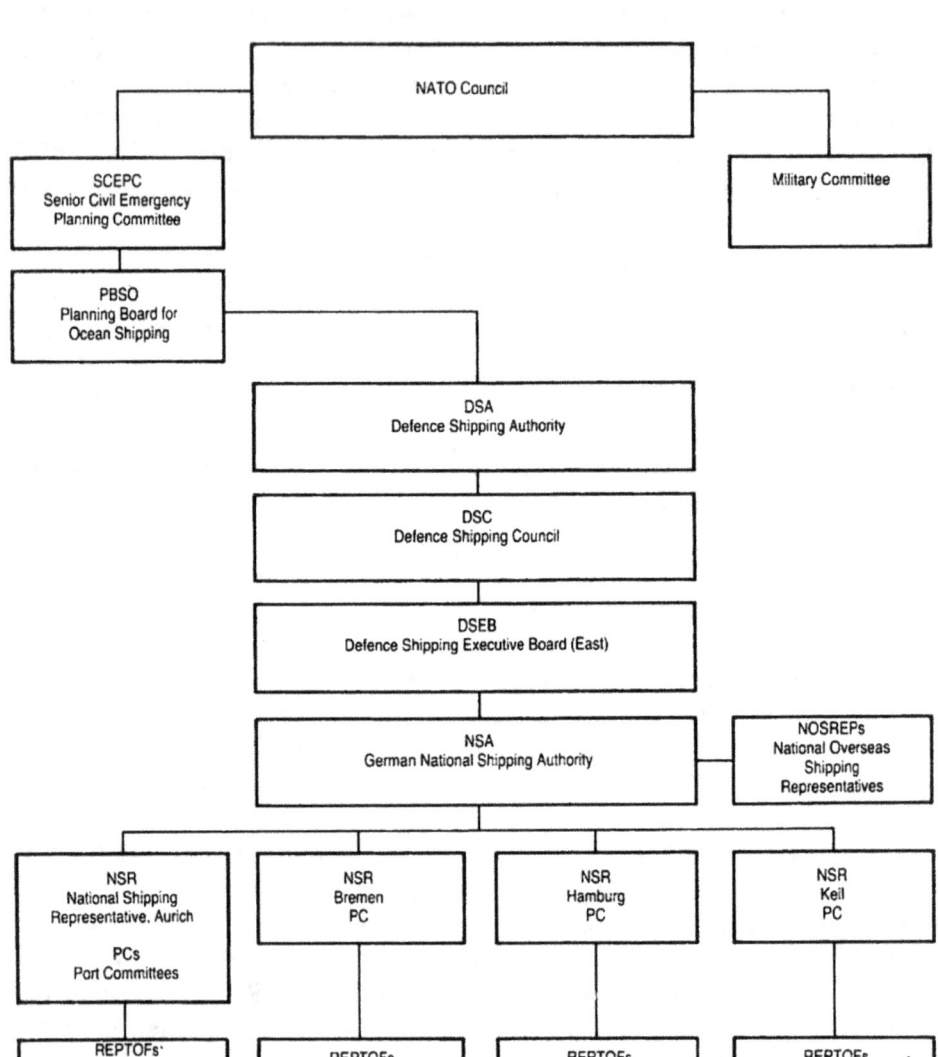

Figure 5.6 Elements in German civilian control of shipping
Scource: adapted from Witthoft (1975)

activating their NCSORGs. In the Baltic and North Sea, the Germans and the Norwegians could be expected to establish MERZONEs.

What would be likely to happen would be that various components of the allied-wide naval control of shipping organization would 'click' in as tensions mount and the Soviet fleet takes to the seas. However reluctant

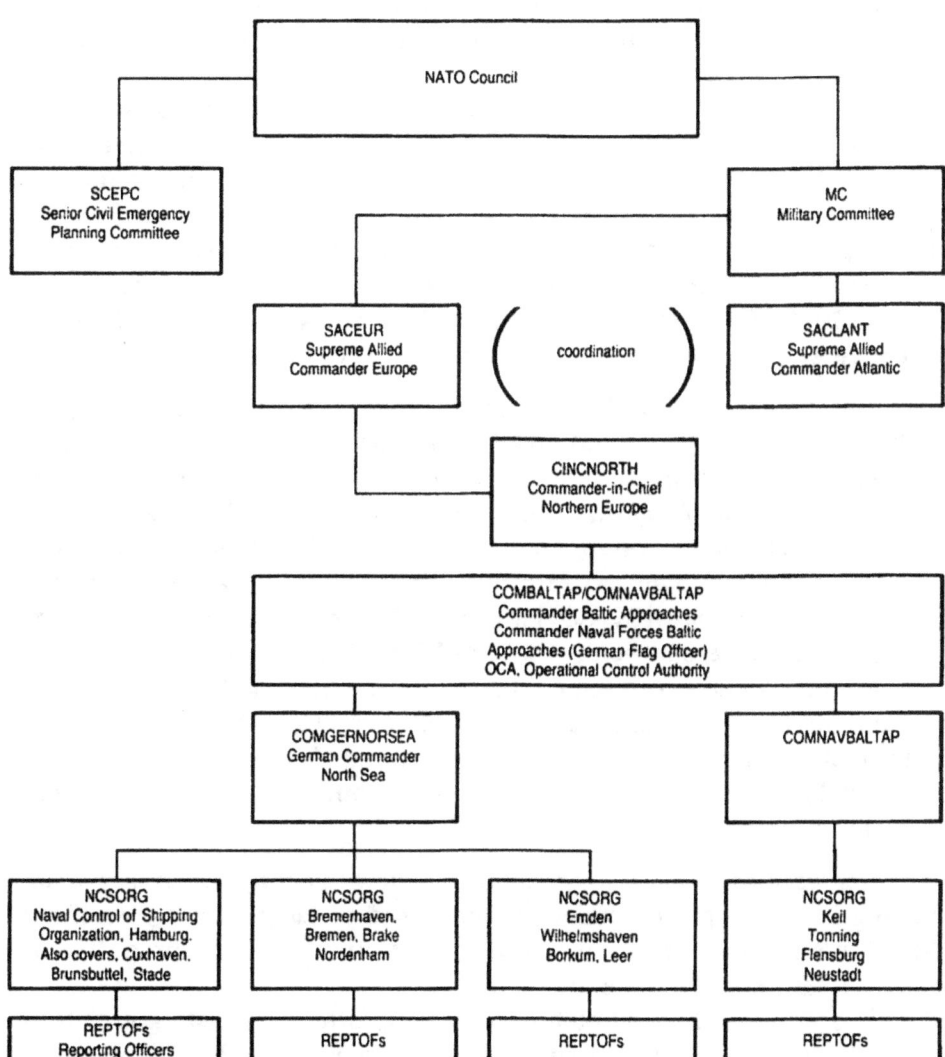

Figure 5.7 Elements in German naval control of shipping

some allies would be to place their shipping under NATO control, the impetus to seek the safety of the combined allied navies, with their attendant intelligence and communications facilities, would be great.

NATO would also be concerned with shipping outside the treaty area. In the past this did not receive much attention, but with the growing range of operations of Soviet submarines there was increased concern with this potential threat. In March 1982, a number of NATO countries, several

Commonwealth countries as well as Brazil and Argentina carried out a major exercise to test shipping control in areas of the world beyond the treaty region (Winton 1983: 340).

Naval control of shipping would be most important for convoy operations: their organization, protection in transit and arrival at destination ports. The first convoys would likely be made up of high-value reinforcement shipping carrying American equipment to US forces in Europe. NATO planning for convoys drew heavily upon the Second World War experience. In a 1973 letter to Admiral Richard Colbert, USN (then NATO CINCSOUTH), a participant in CINCLANT's Convoy Commodore Course (a retired US Rear Admiral) observed that:

> Graduation from the Convoy Commodore Course left me with the distinct feeling that should mobilization be my lot, I would be taking to sea doctrine generated in World War II against an adversary prepared for World War III.... Since the organization of NATO the Navy has approached the convoy mission and related problems on the basis of execution of plans to reinforce Europe at the earliest after the outbreak of war with the Soviet Union. Convoy tactics presented to us in the course were basically those of World War II, modified in places to counter a nuclear threat or the submarine launched surface-to-surface missile. As a result, prospective Convoy Commodores are being trained only for billets in possibly an outmoded type of war, where the resupply of Europe against the Soviet submarine threat is the only scenario.[27]

The Rear Admiral went on to complain that he was being given inadequate training in new types of convoys, composed of larger, faster merchant ships traveling with fewer escort destroyers and facing a 'three dimensional threat' of surface, air and submarine.

This critique of NATO planning was one often directed against SACLANT: the Alliance was intending to fight the battle of the Atlantic over again. In the most general sense, this was true. In the event of war, NATO would seek to bring reinforcements by sea to Europe and these ships could well face a significant submarine threat, particularly as they approached the European coasts. In a high-threat maritime environment, especially when the slower break-bulk cargo ships were being used for the initial unit deployment, convoys would be employed to maximize protection. They would also be used for critical civilian shipping, particularly oil deliveries. In 1973, the Joint Chiefs of Staff issued another version of its *Allied Guide to Masters* (ACP 148[b]) to all allied merchant ship masters. The guide outlined steps to be taken in the event of war, naval control of shipping procedures and particularly convoy organization and tactics. It also set forth basic defensive measures to be taken in the event of air, submarine and surface attack, including missile and nuclear attacks (US Dept of Defense 1973: Chap. VII).

During the 1970s, the US and NATO began to reintroduce actual (as opposed to paper command post) convoy exercises. No major convoy exercise was held between 1964 amd 1972 (Granger 1977: 4). These new exercises were held in conjunction with the annual NATO REFORGER (Return of Forces to Germany) exercises. For example, during REFORGER 1980, 15,000 M/T of US Army equipment was assembled at the Beaumont, Texas port from all over the US. Included was 1,700 separate pieces of equipment, 1,300 of which were tracked vehicles to be used to support and sustain the 2nd Armored Division. Troops from the Division were airlifted (along with small vehicles) to take up prepositioned equipment in Germany. Military Sealift Command ships, both break-bulk and RO/RO, took the equipment from the US to two European ports in Belgium and the Netherlands. Along the way, ships from the NATO navies exercised convoy protection tactics.[28] This was a mock initial reinforcement lift. Former SACLANT, Admiral Isaac Kidd, testified before Congressional committees that a full-scale reinforcement effort, with follow-on resupply and support of European economies at 50 per cent of their pre-war levels, might require as many as 13 to 15 convoys at sea at one time, each composed of 70 vessels (US Congress 1980b: 155).

During the course of NATO's exercise STRONG EXPRESS (1972), involving a conventional conflict along the northern flank, six convoys were organized as part of the testing of allied capabilities. An eastern Atlantic convoy of 13 merchant ships sailed from Portugal to the English Channel; a UK south coast convoy sailed from Torbay to ports in the Bristol Channel; a UK east coast convoy sailed from Loch Eve in Scotland to the Tyne; a North Sea convoy of 36 ships sailed from ports in Denmark and Germany, making a circular passage of the sea and exiting through the Baltic approaches; a German military convoy of 16 ships also crisscrossed the North Sea; a Western Atlantic convoy of 10 ships sailed to Europe from Halifax and Boston escorted by Canadian and American destroyers. A NATO-wide naval control of shipping organization provided convoy operational briefings throughout the exercises.[29]

While NATO would utilize large convoys to sustain forces in Europe in the event of a prolonged conventional conflict, an attempt would be made to provide for sealift and protection of shipping in advance of the organization and transit of large convoys. During a crisis SACLANT, and the US MSC, might decide to send reinforcements in fast-moving ships sailing in pairs or individually before the seas were cleared. Container ships could be modified to carry wheeled equipment and use could be made of large oil tankers. With over 6,000 ships crossing the Atlantic daily, it would be difficult for the Soviets to locate these ships. Even if they did find them, it would be unlikely that they would sink individual ships during a crisis period because of the danger of escalation. Hostilities, therefore, could break out while these ships were in transit and they would become subject to attack. In anticipation of this, NATO in the late 1970s began testing procedures for the protection of individual, high-value ships. During Exercise OCEAN SAFARI in 1977, several merchant ships were requested to change their routes so that allied naval vessels could test procedures for the protection of unconvoyed ships converging on the southern approaches to the English Channel (Wettern 1977: 4).

As the level of tension grew, and as the Soviets moved more anti-shipping submarines, aircraft and surface vessels out to sea, SACLANT would order the seas cleared. Reinforcement shipping would then form into convoys 'at designated assembly areas off the east coast of the United States' and proceed to Europe over a preselected route (US Dept of Defense 1976: VIII–8). Depending on the success of the allied ASW campaign in the north, it might be necessary for reinforcement shipping to take a more southerly route. The growing threat of Soviet long-range naval aviation could also compel SACLANT to route convoys further south towards Maderia, for example, turning northward as they approached the continent.

Another problem for convoys was the range of new, anti-ship missiles. These allowed enemy forces to strike at ships in transit from distances beyond the range of the sensors and ASW weapons of close escorts. For this reason NATO began examining an old concept, Defended Lanes. Originating in the years prior to the First World War, this notion sought to protect merchant shipping by having surface ships patrol the routes traveled by individual merchant ships, rather than forming the ships into convoys with close escort protection. In both World Wars the initial heavy losses persuaded the British Admiralty to abandon this approach and return to the convoy system (see Winton 1983).

With modern weapons now capable of striking at convoys from distances of up to 250 miles, there is a need to provide extended defense with helicopters, submarines and long-range patrol aircraft. In NATO's 1981 OCEAN SAFARI exercises, this method was tested on mock convoys moving from Lisbon to the English Channel. The results were mixed, indicating that, under intensive submarine pressure, convoys with close escort as well as extended protection by ASW forces will be needed. To fulfill the latter requirement, NATO provided funds for the development of a base for maritime patrol aircraft at Porto Santo in the Maderia group of islands (Wettern 1981: 718, 722).

Situation	Organization	Activity
Peace	SACLANT, NSWG and SACEUR	Planning
Tension	Activiation of national NCSORGs and some NATO shipping control, US NCSORGs under CINCLANT	Establish MERZONEs, watch individual ships, some sealift sailings
War	NATO-wide NCSORGs under OCAs	Convoys and defended sealanes

Figure 5.8 Sequence of NATO naval control of shipping

NATO plans for naval control of shipping cannot be dismissed as an example of old Admirals preparing to fight the last war over again. As noted above, to the extent that the Alliance might need reinforcements from the US, of a magnitude which could be moved only by sealift, in order to support a flexible response posture, the control of shipping problems had many similarities to past experiences. Ships would have to sail through waters in which enemy submarines, surface ships and naval aircraft were operating to reach ports and bring supplies to ground forces. The use of tactical nuclear weapons at sea against convoys posed an unprecedented threat forcing greater distances between ships. However, a 40-ship convoy, with 5 miles between each ship, would cover an area of 750 square miles, making it difficult for even several nuclear weapons to destroy the entire group (Winton 1983: 344).

To be sure, a rapid escalation of nuclear combat, during which port facilities and ships were subjected to massive nuclear strikes, would deprive sealift of much of its utility. However, in the event of an initial nuclear exchange being confined to Europe, with the war continuing afterwards, the US may well have to lift enormous amounts of supplies to a devastated population. With a continuing conflict, this sealift could still be subject to Soviet attack. In naval control of shipping, as in all other aspects of the Alliance's maritime posture, planning was based on the assumption that allied maritime forces must be prepared to support as full a range of contingencies and responses as resources would allow.

Moreover, it is evident that certain adjustments were made to modern conditions. For example, the very process of on-going planning by an allied body reflected an appreciation that, in the next European war, allied-wide naval control of shipping would have to be established quickly, preferably before the outbreak of hostilities. Plans regarding the initial movement of reinforcements in non-convoyed ships also reflected a realistic appreciation of the differences between a future NATO war and the Second World War.

NATO REINFORCEMENT SEALIFT: AN ASSESSMENT

No other aspect of NATO's maritime posture more vividly demonstrated the continuity between the strategic use of the sea in the pre-nuclear age and in the post-1945 era than the Alliance's efforts to develop plans and procedures for a reinforcement sealift. It was the ability to transport troops and equipment across the seas which allowed Rome to defeat Carthage, Britain to defeat Napoleon and the Second World War allies to defeat the Axis powers. Since the beginning of NATO, the United States and its allies have sought to counter the threat posed by the presence of large Soviet land armies in Europe. Thus, amongst the myriad other measures taken in the name of collective deterrence, the Alliance attempted to provide the means to execute a reinforcement sealift of forces which would help blunt a Soviet land assault. As before, it was the ultimate strategic and political importance of land power, of the defense of definite frontiers, which gave seapower its relevance.

Concern with the timely and adequate availability of sealift capacity was a continuing theme in the evolution of NATO's maritime strategy. The importance of sealift during the early years, with the Alliance's emphasis on nuclear weapons and massive retaliation, was not particularly crucial. Nevertheless, PBOS met regularly and, in the US, periodic assessments were made of the ability of the United States to execute a reinforcement sealift, with or without NATO flag shipping.

It was in the late 1960s and early 1970s that sealift increasingly gained importance in the context of NATO's maritime and overall strategic posture – a posture which placed more emphasis upon conventional deterrence.

An assessment of the wisdom of the efforts taken by NATO to improve its sealift posture in support of conventional deterrence must begin with some reference to one of the key assumptions that underlay this improvement. The assumption was that the Alliance would have sufficient strategic warning of an impending Warsaw Pact attack, to at least commence a reinforcement sealift from the United States. Estimates as to how much time would be required between the moment of mobilization and the outbreak of war varied, but there was general agreement that the process had to begin as early as possible. Ideally, reinforcements would arrive by ship before the outbreak of war on land and before sea lanes of communication came under attack.

Critics of this assumption were numerous. Many, such as Maxwell Taylor, former Chairman of the US Joint Chiefs of Staff, argued that the Soviets could launch a short-warning attack which could overwhelm NATO forces 'after a short-period of intense conventional conflict'. Thus no program to reinforce NATO, especially sealift, could be considered 'adequate' (Mako 1983: 75).

The Alliance responded to the threat of the enhanced capacity of the Warsaw Pact to launch a short-warning attack, through efforts to improve rapid reinforcement, particularly with more prepositioning of equipment for American forces. But there were limits to how extensively the Alliance could prepare for a short-warning attack. These limits were those not only of adequate resources, i.e. sufficient prepositioning and airlift, but of allocation of available resources and overall prudence. As noted above, prepositioning and airlift could not guarantee the levels of reinforcement that were required to sustain conventional resistance. Thus sealift, the acquisition of available capacity, its organization and protection would be essential. Beyond this it was evident that, without proper planning to execute a sealift, NATO would not be able to use a period of strategic warning effectively should one occur. That is, if the critics were wrong, if a war in Europe emerged out of a period of rising tensions, if the Soviets did mobilize and NATO responded with a counter-mobilization, it would be important that peacetime planning for a reinforcement sealift be in place.

According to William Kauffman, the Alliance was placing too much emphasis upon the short-warning scenario:

> Instead of betting so much on the relatively untested and difficult process of airlift and prepositioning – adopted to a considerable degree because of the exaggerated estimates of the Soviet ability to launch a surprise attack with conventional forces – cheaper more reliable sealift could be given greater consideration for most of the US reinforcement capability.
>
> (Kaufman 1981: 173)

The Alliance attempted to provide both for a short-warning response and for the possibility of a period of mobilization sufficient to allow for sealift of reinforcement. Given the uncertainties that attended all allied planning, this did not seem unreasonable. To be sure, even if NATO could be more certain that it would be able to begin a sealift, there would still be significant problems, not all directly related to the ships themselves.

There was, as well, the whole problem of American readiness. For example, in 1979 the US Defense Department and civilian agencies conducted NIFTY NUGGET, which was the first full-scale mobilization test since the Second World War. Although only a paper exercise, it showed that, while adequate capacity was readily available for an initial European sealift, there were serious problems in getting ships to the right ports. Other problems included difficulties in moving equipment to loading ports. Still further, it was found that to resupply US forces already deployed, equipment had to be taken from units scheduled for movement during the war's second month. In general, there was considerable confusion about the steps necessary in an American mobilization (US Dept of Defense 1980: 17).

On the other side of the Atlantic, there was concern that the ports and unloading facilities upon which the sealift depends would not be available. In a resupply of the forces deployed in Germany, the German ports are particularly useful not only because of their proximity to the front, but also because unloading there allows for reinforcements to move in without crossing operational lines.[30] Yet, in the event of war, these ports might be under air attack and the object of mining. Thus, the success of the sealift effort would depend upon both the outcome of the war at sea and especially upon the progress of the land and air war. Indeed, the ability of the entire allied maritime organization to make a flexible response possible depended heavily on the initial character and result of the land and air war as well as the efficacy of unloading and internal transportation in Europe.

Here again, however, efforts have been made to improve the allied logistical posture. After the NIFTY NUGGET exercise, the United States created a new Joint Deployment Agency. This agency was 'tasked with co-ordinating the planning of all multiservice military deployments and with developing an automated date management system to support future deployment actions'. In order to improve off-loading in Europe, the US entered into a new series of bilateral Host Nation Support agreements (HNS) with Britain, Belgium, Holland and Luxembourg. Under these HNS, the host nations will supply additional manpower, such as German territorial reservists, to move equipment from debarkation ports to military assembly areas (Simons 1984: 19–21).

As to the availability of shipping capacity, there were two broad problem areas: whether the right kind of ships will be available, and – the more crucial consideration – will the allied governments release them to the US military in time?

One of the first problems, the difficulties posed by increasing containerization, has already been outlined above. In general, the allies did not maintain their merchant fleets with a view toward their possible use in war. However, it would have been somewhat unrealistic to expect otherwise. No nation, not even the United States, could afford to keep in reserve, in a high state of readiness, the kinds and amounts of shipping required to sustain fully conventional resistance in Europe. And, while all allies, including the United States, subsidized their merchant marine for military and economic reasons, it was not possible for the Europeans to subsidize a merchant marine for military reasons alone. As H. G. Davy, Director for Defence Affairs of the General Council of British Shipping observed, 'the allies cannot afford to sustain a merchant fleet of the size and composition entirely suitable to military or other maritime war requirements' (Davy 1982: 67).

Yet the problems posed in this regard were not insurmountable. First, the United States moved forward with the development of a fast sealift ship, the SL-7. Second, as the British showed during the Falklands War, commercial ships cold be made ready for war in a short period of time. In all, the British took some forty-five ships from commercial fleets, extensively refitting some, and used them to support their forces at great distance. Liners, such as the *Queen Elizabeth II* and the *Canberra*, as well as ferries, were used as troop carriers. Cargo ships, including break-bulk, were employed, while a container ship was employed to carry helicopters, Harriers and heavy equipment. Although there were problems in unloading due to the lack of adequate port facilities, the acquisition and deployment of civilian sealift capacity represented a 'smooth and rapid implementation of existing contingency plans to use merchant shipping in support of the Services'.[31]

The second broad problem, that of the willingness of the allied governments to make adequate capacity available early enough, was of continuing concern to the United States. It was this concern which prompted the US to seek a greater pre-allocation of allied civilian capacity during the 1970s. While no definitive answer can be given as to whether this capacity would be available in a crisis, a number of considerations should be kept in mind.

Viewed from a comparative perspective, it is evident that the question of allied response to a reinforcement sealift was not unique. Doubts as to how NATO would respond collectively to a Soviet attack or threat of attack pervaded the whole panoply of allied plans and organizations. Yet, the Alliance went forth with these doubts. The implicit hope was that the Soviet leadership would be sufficiently convinced of the possibility of a strong unified response that Moscow would refrain from attacking or threatening to attack.

The shipping arrangements examined in this chapter were part of the whole range of plans and organizations which contributed to collective defense. Unlike the Alliance's nuclear forces and conventional elements, most sealift capacity was not under military control and ready for battle in peacetime. This did not mean that a decision to employ military force would therefore be more readily forthcoming than one to make shipping available. Indeed, it can be argued that, although an escalatory step, a collective decision to begin the process of allied control of shipping for reinforcement purposes would be more likely to be made than other decisions. A shipping decision would be likely to occur within the context of a growing crisis and represent a controlled, prudent response. The real problems in obtaining allied unity would develop once hostilities break out, for example over the use of tactical nuclear weapons.

There remained, however, grounds for concern on the part of the United States that its European allies would withhold all or some of their shipping assets. The allies might be worried about maintaining an adequate level of civilian shipping to support their economies, even during a period of tension. More significantly, they might not wish to support a sealift operation for fear of its escalatory impact and possible Soviet reaction. The Europeans might decide to initiate an airlift and await developments. In a worst case scenario, the Soviets attack without warning, across the central front or the flanks. NATO retreats, giving up territory. Rather than use nuclear weapons, the Europeans decide to meet Soviet demands. Under these circumstances not only would sealift be ineffective, but the European allies would not request reinforcements.

To take the first rationale for withholding, it must be pointed out that the initial requirements for a reinforcement sealift during a period of tension represented less than 10 per cent of the total allied shipping normally at sea. So, making earmarked ships available to the US would not seriously strain allied economies. In the event of a prolonged sealift operation, possibly in the face of Soviet anti-shipping attacks, withholding of ships may harm the reinforcement effort. By this time, the DSA would have been established to pool all shipping, civilian and military. In addition, Alliance-wide naval control of shipping would have been established. While it is possible that in a lengthy conventional confrontation some allies might withhold shipping for civilian purposes, it does seem unlikely that they would frustrate a reinforcement effort, essential to their own protection, in the middle of a war. After all, what value would it be to support only civilian needs at the expense of exposing the civilian population to greater physical danger?

As to the second rationale for withholding shipping – that is, for one reason or another the European allies did not wish to begin a reinforcement sealift because of its escalatory effect – again, this was not beyond the realms of possibility. Moreover, a delay or indecision on the part of the Europeans would, as noted, undermine the effectiveness of a sealift executed in conjunction with airlifting US troops to prepositioned equipment.

But the key decision would be the one to request reinforcements, not the one to supply shipping. A reinforcement sealift would be initiated during a crisis period if the North Atlantic Council decided to bolster NATO's position on land in Europe, in order to demonstrate to the Soviets that the Alliance was prepared to meet an attack at all levels and sustain that resistance. In the event of a short-warning Soviet assault, it is likely that the Council would issue a request for reinforcements even as it contemplated the use of nuclear weapons. In either case, it would be illogical for the allies, especially the Germans and those on the flanks, to request reinforcements either individually or collectively and then withhold the shipping necessary to lift the bulk of the equipment and resupply.

The United States could begin reinforcement without waiting for a request from the allies. During a prolonged period of tension, the US would probably move unilaterally to strengthen its forces in Europe. Steps would be taken to break-out reserve shipping and, following a presidential declaration of national emergency, to requisition American commercial vessels. It is unlikely, however, that the President would order a major sealift unless the allies, or significant numbers of them, both requested reinforcement and were willing to supply the necessary capacity to undertake a major reinforcement within the timeframe required to make it effective. In the first place, without rapid augmentation of US shipping by NATO flag vessels, a major sealift could not take place. In the second, if the allies could not agree on the need to provide ships, then it is probable that other major disagreements existed as to how to respond. A situation in which sealift capacity was being deliberately withheld would be one of allied disunity. Under these circumstances, a major American sealift reinforcement effort would not commence immediately, if at all, and the question of European shipping availability would be temporarily moot.

As the SPANS study noted, the European allies would make shipping available to the US for reinforcement purposes at levels commensurate with their estimate of the severity of the situation. The December 1982 ministerial approval of an Alliance-wide plan for crisis reinforcement in Europe, coupled with other improvements in the procedures for an early acquisition of sealift capacity by the US, were a strong indication of the broad acceptance by the allies of the need for a credible sealift posture to bolster deterrence. In the final analysis, the commitment of the European allies to supply ships for a reinforcement sealift was as reliable or unreliable 'as the commitment of those countries to the common defense. As they will fight to defend themselves, they will provide ships to transport reinforcements for the battle' (US Congress 1980a: 90).

Chapter 6
The USN, NATO and the war at sea

ASSESSMENT OF THE RISK

The *Brosio Study* conducted by SACLANT in 1969 had warned of a serious threat to the allied capacity to secure the use of the seas by the mid to late 1970s. Assessments conducted in subsequent years by the United States Navy had the same theme. NATO was only marginally superior to the Warsaw Pact at sea and this thin edge was quickly vanishing. To be sure, there was a certain amount of self-serving pessimism when such assessments were conducted by naval officers. Invariably the solution proposed was more expenditure on maritime forces. However, even allowing for service interests, the trend of assessments done throughout the 1970s and into the 1980s suggested that there was less confidence about the capacity of allied maritime forces to win a conventional war at sea in a manner and within a timeframe acceptable to NATO. These assessments also raised questions about the ability of the Alliance to carry out successfully the forward operations and sustained initiatives contained in the 1980 Concept of Maritime Operations.

In the spring of 1973, the US Navy's Office of the Chief of Naval Operations evaluated the American and allied maritime forces' capability to carry out their missions in the context of a large-scale non-nuclear war. According to subsequent Congressional testimony by CNO Elmo Zumwalt, 'the primary mission of these forces in such a situation, is control of the seas in the vicinity of our vital sea lines of communication and in key shore areas where such control is essential to the conduct of the war'. Assessments were conducted for forces available in 1974 and those which would be available in 1981 (if Congress approved the Navy's plan). Winning did not entail destruction of the enemy's forces, rather the ability of the combined American and allied forces to protect shipping 'necessary to resupply and reinforce the US and allied forces conducting a conventional defense' and to protect and secure the northern and southern flanks 'while the land war is fought in the center'.[1]

In summarizing the results of the assessments, Zumwalt indicated that the 'present' (1974) situation raised questions abut NATO's ability to fulfill its maritime missions. Stressing that 'absolute quantitative measures' could not fully define success of the NATO navies in fulfilling their objectives of protecting the sea lanes and securing the flanks, he told a Senate Committee that:

> [O]ur analysis, our net assessment, the result of our Fleet exercises, the judgement of our Fleet commanders and my own personal judgement, indicate that the odds are that we would be unable today to succeed in these goals. If we fail to succeed in these goals, the odds are that we would lose central Europe and the northern flank countries.[2]

At the time, Zumwalt himself rated the chances of winning in the NATO context at only 30 per cent. Acceptance by the Congress of the full Navy program would, according to Zumwalt, have improved the odds to 50–50 by the 1980s (Zumwalt 1976: 64). This Navy assessment was supported by a study initiated in the fall of 1973 by the President's Foreign Intelligence Advisory Board (PFIAB), based upon information gathered independent of naval channels. Completed in early 1974, this study offered an even more pessimistic evaluation than that of the Navy's studies (ibid.: 465).

These assessments had the US and allies undertaking all missions simultaneously. In the context of an evolving conventional conflict, wherein the central front is the focus of hostilities, NATO might, with greater chances of success, confine its initial actions to securing sea control for reinforcement purposes and holding off on projection. This, at least, was the conclusion of an assessment undertaken by the Congressional Budget Office (US CBO 1976).

The CBO study argued that the US, with the help of its European allies, would be able to secure control of the seas for the purposes of reinforcement and resupply. This would entail a conflict of several weeks, perhaps two months, and involve heavy losses to both naval units and merchant shipping, but, in the end, NATO would win. In attempting to secure the SLOC, the major threat would be the Soviet submarine force. Other elements of the Soviet Navy, surface ships and naval aviation, would constitute less of a threat because of the distances they would have to cover to mount a concentrated attack on allied shipping (ibid.: 24).

Projecting force ashore would require that NATO carrier task forces move close to the Soviet Union. Within a range of 500 kilometers, these forces would come under concentrated attack, for example in the Norwegian Sea and eastern Mediterranean. Under these circumstances:

> Combined Soviet systems are likely to exert enough pressure on naval forces to force them to be preoccupied with their own survival rather than with the projection of power ashore.
>
> There are serious doubts that even self-defense would be successful given the intensity of attacks that Soviet shore and medium range defenses can mount. Finally, attack of such Soviet targets as can be reached from the sea would not be likely to affect the outcome of the war in central Europe significantly.
>
> (ibid.: XV, 12)

This general assessment of the higher risks associated with projection, as compared to securing the SLOC, was expressed in US Defense Department Congressional testimony in the later 1970s. In 1977, Secretary of Defense Harold Brown stated that NATO had enough forces to secure the sea lanes in order to allow for reinforcement, while acknowledging the threat to carrier forces in the eastern Mediterranean.[3] Testifying a year later, the Deputy Secretary of Defense noted that: 'For the present, we believe that the US Navy can carry out the sea control mission in concert with our allies.'[4]

The US Navy's view was similar, but more guarded. According to CNO James Holloway, 'the US could probably retain control of the North Atlantic sea lanes to Europe, but would suffer serious losses to both US and allied shipping in the early stages' of a conflict. He put the ability of allied forces, particularly the Sixth Fleet, to operate in the eastern Mediterranean as 'uncertain at best'.[5]

The degree of risk inherent in these assessments was considerable. First, regarding the danger of attempting to project force along the flanks, it should be noted that, even if NATO did not strike at targets within the Soviet Union (and this was not the primary mission of carrier forces), the concentrated firepower of the Soviets could have prevented effective support of the land battle. Under flexible response, all NATO territory was to be defended initially with conventional forces. In the eastern Mediterranean and in northern Norway, projection forces were to play an important role in this defense. Moreover, these flanks were viewed as the most likely areas of limited Soviet attacks because of the risk of nuclear escalation along the central front. Clearly, a limited, and appropriate, NATO response in the event of such an attack was becoming more and more problematic.

The second point relates to the implication for the wider sea battle of a loss of the flanks to conventional land forces. Had the Soviets won the air and land war near the Turkish straits or in northern Norway, they would have been able to move their maritime aircover forward and provide a more secure environment for other maritime forces. In the north, the combination of a successful strike against allied carriers and advances on the ground could have resulted in NATO losing the battle of the Norwegian Sea. According to one SACLANT, 'control of the Norwegian Sea by NATO forces is a vital part of allied defences against Soviet submarines, since the level of threat posed to the Atlantic SLOC will be a function of the Norwegian Sea battle outcome' (North Atlantic Assembly 1982: 25).

This, in turn, led one naval analyst to point out that:

> A war between NATO and the Warsaw Pact cannot be won either at sea or on the flanks. It must be won on the primary battlefield, the central front. But who wins in the center and the ability of the winner to realize the benefits of victory could both be determined by events on and around the northern flank. Either side could lose the battle that it must fight there and thereby lose the war.
>
> (Weinland 1980: 2)

The third point is that, while the Soviets might have been prevented from mounting concentrated surface, subsurface and air attacks on convoys and allied maritime forces below the GIUK gap and the western Mediterranean, the same situation would not necessarily hold true in the coastal waters and harbor areas of northern Europe. In a major war in Europe, the most important transatlantic military and economic shipping would converge on the English Channel and pass through to the North Sea ports of Belgium, the Netherlands and Germany. 'Given the opportunity, the Warsaw Pact could have been expected to have concentrated both submarine and air attacks in those waters.' The Channel sea lanes and North Sea ports would also be 'excellent sites for Soviet minefields' (US CBO 1980: 48–9). With their surface ship and air-launched mines, the Soviets were 'able to recover all the northern European sea lanes' (Kidd 1980: 200). Thus, short-range Soviet capabilities would also have put into question the Alliance's ability to land reinforcements, regardless of the level of sea control obtained through combat, on the open seas.

Such assessments indicated serious concern over NATO's ability to provide maritime support in the event of a conventional land and air war in Europe. They did not (and most allied naval commanders stressed this) prove that the Alliance would lose the war at sea in the 1970s. The situation was uncertain, not hopeless. And in this sense, such assessment did not differ substantially from assessments of other aspects of the NATO posture. Moreover, it is important to point out that there were solutions to some of the problems the allied maritime forces might face. For example, the vulnerability of carriers could be compensated for through greater reliance on land-based aviation along the northern and southern flanks (see Zakheim 1978b and 1982). Transatlantic convoys could be routed well beyond the range of Soviet naval aviation and at distances which would put Soviet submarines attempting to cut the SLOC in considerable danger. From the standpoint of SACEUR, it was preferable if reinforcements and resupply could come in through northern European ports, and British reinforcements move across the Channel. However, depending on the situation with France, it would have been possible to use French ports.[6]

In the Mediterranean, the continued presence of the US Sixth Fleet provided a large measure of assurance that at least NATO would have a major force in the region from the outset of hostilities. There was a high degree of confidence that the Soviet surface threat could be 'neutralized'. Soviet naval aviation would be a threat, as would submarines, although NATO could have mounted effective ASW barriers to prevent large numbers of submarines from entering the Mediterranean after the outbreak of hostilities.[7]

One assumption which tended to raise the level of confidence that NATO maritime forces would be able to at least secure the SLOC, if not project force ashore, was that of sufficient strategic warning time. Here again, maritime assessments were similar to those often made with regard to NATO conventional land and air posture. The balance of maritime forces, only marginally favorable to NATO, would improve if the Alliance was able to begin mobilization and deployment of its forces well in advance of hostilities. This would mean the establishment of ASW barriers particularly along the GIUK gap, commencement of intensive area searches and the movement of US carriers toward Europe. It would also have involved the deployment of minesweepers in the Channel and harbor ports, as well as the mobilization of land, air and maritime forces under Commander Baltic Approaches.

As NATO began to deploy its maritime forces, however, the Soviets might also have been moving theirs out to sea. Under strict rules of engagement, imposed by higher political authorities still seeking a negotiated solution that would avoid war, allied maritime commanders could only have tracked Soviet movements. Thus, although early warning and deployment was considered necessary if NATO was to win a conventional war at sea, long warning times also raised the prospect of early major engagements. As former SACLANT Admiral Harry Train II, USN, told

Table 6.1 Comparison NATO and Warsaw Pact naval combatants 1983 (1)

	BBs, Cs, DDs, FFs(2)		CARRIERS (3)		SUBMARINES (4)		TOTAL		RESERVE UNITS (5)		TOTAL ACTIVE & RESERVE UNITS	
ACTIVE UNITS	NO.	TONNAGE	NO.	TONNAGE	NO.	TONNAGE	NO.	TONNAGE	NO.	TONNAGE	NO.	TONNAGE
BE	4	9,132	0	–	0	–	4	9,132	0	–	4	9,132
CA	20	66,226	0	–	3	7,230	23	73,456	3	8,640	26	82,096
DE	10	17,970	0	–	5	2,829	15	20,799	0	–	15	20,799
N FR	45	135,535	3	77,925	24	75,122	72	288,582	0	–	72	288,582
GE	15	56,680	0	–	24	11,664	39	68,344	0	–	39	68,344
GR	20	60,980	0	–	10	14,895	30	75,875	0	–	30	75,875
A IT	22	70,642	1	8,850	10	14,252	33	93,744	0	–	33	93,744
NL	18	61,326	0	–	6	12,704	24	74,030	0	–	24	74,030
NO	5	8,725	0	–	14	6,090	19	14,815	0	–	19	14,815
T PO	17	28,542	0	–	3	3,129	20	31,671	0	–	20	31,671
SP	26	70,834	1	16,416	8	12,512	35	99,762	0	–	35	99,762
TU	17	54,575	0	–	15	30,605	32	85,180	0	–	32	85,180
O UK	52	184,060	3	67,700	31	125,750	86	377,510	3	8,100	89	385,610
US	190	1,154,898	13	1,045,461	133	815,581	336	3,015,940	43	574,454	379	3,590,394
TOTAL	461	1,980,125	21	1,216,352	286	1,132,363	768	4,328,840	49	591,194	817	4,920,034
W BULG	2	2,640	0	–	2	3,600	4	6,240	0	–	4	6,240
P E.GERM	11	14,800	0	–	0	–	11	14,800	0	–	11	14,800
A POLAND	1	3,600	0	–	4	5,400	5	9,000	0	–	5	9,000
C USSR	296	1,090,770	5	160,000	378	1,854,800	679	3,105,570	108	176,900	787	3,282,470
T TOTAL	310	1,111,810	5	160,000	384	1,863,800	699	3,135,610	108	176,900	807	3,312,510

NOTES: (1) Based on *Jane's Fighting Ships 1983–1984*, figures for end December 1983
(2) Includes vessels over 1,000 tons full load only. Excludes amphibious ships.
(3) Includes CV(Helo)s but excludes Commando-type carriers (e.g. US LHA/LPH).
(4) Includes SSBNs and SSBs (FR–6; UK–4; US–34; USSR–84).
(5) CA–3FF; UK–3FF; US–2BB, 2CG, 2CA, 19DD, 6FF, 5CVA/CVS, 6SSN, 1SS; USSR–13DD, 10FF, 85SS.

a Congressional committee in 1981: 'The greatest area of uncertainty in assessing the maritime situation in the Atlantic is the transition from peacetime operations to crisis management to war – sometimes referred to as the "D-day shoot".'[8]

The prospect of early, major engagements at sea in the context of a still conventional land and air war further added to the uncertainty and risk facing allied maritime forces. The D-day shoot-out could prove inconclusive, only the first round in a more protracted war of attrition at sea which takes a heavy toll of both NATO and Soviet maritime forces. This, in turn, raised the possibility that the war-at-sea might be won too late – too late to affect the course of the land and air war, at least on the conventional level.

Along with the assumption that allied maritime forces would be at sea on D-day, NATO also counted on a pre-hostilities commencement of the reinforcement sealift. While necessary, this early movement would mean that merchant ships, whether sailing independently or in convoy, would have been immediately vulnerable since they would have been at sea before NATO was fully able to secure the SLOC. Much would have depended on the disposition and intentions of the Soviet fleet. If the Soviets held back their forces to protect their SSBN fleet and did not enter the mid-Atlantic, NATO ships at sea might well have proceeded safely behind an ASW barrier along the GIUK gap and behind the battle in the Norwegian Sea. On the other hand, if the Soviets had moved out in significant numbers in advance of hostilities, not only would the battle to secure the SLOC have been harder, but shipping would have quickly been vulnerable from D-day on.

Estimates of shipping losses varied with assumptions as to how much warning time NATO might have and the extent to which the Soviets would move their submarines out. Former SACLANT Isaac Kidd predicted that 'well over a third and probably more of the merchant ships at sea would be destroyed or prevented from delivering their cargoes on the day the shooting starts' (Kidd 1980: 199). A study by the US Atlantic Council, under the direction of former Secretary of the Navy Paul Nitze, estimated that between 300 and 600 allied merchant ships and escorts would be lost within the first 4 to 12 weeks of a major war at sea. This study, which was said to have been based on statistics supplied by the US Navy's Center for Naval Analyses, assumed a pre-deployment of allied maritime forces and Soviet anti-shipping forces in forward positions (Nitze *et al.* 1979: 374, 381). Another analysis, undertaken by the Carnegie Endowment, suggested that if the Soviets moved significant submarine forces beyond the GIUK gap, it might take a month for the SLOC to be fully secure

(Carnegie 1981: 126). Most projections of relative maritime strength gave the long-run edge to NATO, so that 'if the conventional NATO defense should be successful for more than a month or so, then the West should be able to regain the use of essential sea lanes at acceptable attrition rates' (Nitze *et al.* 1979: 381).

The problem was that NATO would have required substantial sealift well before a month was out and could not wait for the outcome of a protracted battle for control of SLOC. Even with a period of pre-hostilities mobilization and commencement of air- and sealift, the tonnages able to get through in time could have been insufficient. A 1982 report issued by the North Atlantic Assembly's military committee stressed that the 'gross tonnage required for concentrated conflict lasting more than a few days would have to come by sea'. Existing stocks could be drawn down 'within as little as 2–3 days' (NAA 1982: 15). In his 1983 book former US Secretary of Defense Harold Brown argued that NATO did not have the capability to fight conventionally 'for a reasonable period (say a month or more)'. The Alliance could not hold onto the inter-German border. According to Brown, 'the Soviets are probably not certain of being able to drive to the Rhine, the North Sea and the English Channel in a month's time.... Their confidence of doing so,' he pointed out, 'probably exceeds the NATO countries' of being able to stop them. That left NATO with an uncomfortable reliance on the threat of nuclear escalation' (Brown 1983: 101–2).

Allied maritime forces and the sealift they could protect were long viewed as a hedge against both early use of nuclear weapons in a war and excessive reliance upon nuclear weapons for deterrence. It was evident, however, that the credibility of this hedge was undermined not only by NATO's decreasing margin of maritime superiority, but also by continuing doubts as to the ability of its conventional land and air forces to sustain conventional resistance in the absence of a major sealift during the first weeks of a war. The war at sea could have been won, yet the Alliance could still have found itself in a difficult position if this victory was achieved after large-scale losses of land forces and territory. In a real sense, the credibility of NATO's conventional maritime posture, as with other aspects of its overall deterrent, had become hostage to the conventional land and air balance along the central front.

THE NUCLEAR DIMENSION

In looking at the war at sea, the possibility that a war in Europe would either not begin or long remain conventional by virtue of Soviet first use could not be excluded. While the bulk of allied maritime forces were postured for greatest effectiveness in a protracted conventional conflict, they would have been crucial in a war which passed the nuclear threshold either at sea or on land.

At sea, the Alliance could not discount the large numbers of Soviet ships, submarines and aircraft capable of delivering nuclear weapons. Still faced with Western numerical superiority, especially in sea-based air and ASW platforms, the Soviets could have resorted to nuclear weapons, particularly if large numbers of allied forces were close to Norway and therefore close to the Kola Peninsula. Even if NATO did not attack Soviet territory, the presence of allied forces in waters close to the Soviet Union might well have been perceived as so threatening as to elicit a nuclear response.[9] The Soviets could also use the SLBMs on their Yankee class submarines, to strike near NATO task forces with nuclear warheads which only had to land near allied ships to devastate them (see Clawson 1980).

There was also the possibility that the Alliance, faced with a Soviet submarine campaign which threatened not only to prevent reinforcements but took a high toll of escort ships, would ask the United States Navy to employ its nuclear ASW weaponry. A situation might have arisen where the forces on the land and air front were in danger of collapse without reinforcements and there was not enough time to engage in a conventional ASW campaign. The benefits of nuclear ASW weapons over conventional weapons derives from their large legal radius. Their use against submerged submarines would allow for successful engagement where localization was not exact. The blast would also tend to nullify countermeasures attempted by the target submarines. The USN had nearly 2,000 ASW torpedo nuclear warheads and depth bombs which could be delivered by submarines, surface ships, land- and carrier-based aircraft and helicopters (Sokolsky 1984: 154).

Nuclear ASW weapons cannot be used in shallow waters or at shallow depths in the vicinity of friendly vessels and are most effective against deep-diving submarines because of greater existing water pressures. As nuclear weapons, their use would raise similar problems of escalation associated with other nuclear weapons. However, the first use of nuclear weapons in an ASW capacity would not have carried the same consequences as the use of tactical or theater weapons against land targets (ibid.: 155). In mid-ocean areas, there would be little fallout or collateral civilian damage. The allies might well have found it easier to reach a collective agreement on first use at sea, as opposed to initiating a nuclear counterattack on land. Nor could unilateral use by the United States be excluded in defense of high-value American ships such as carriers. Such a step would again have been more likely at sea than along the central front.

The use of tactical or theater nuclear weapons by either side along the central front would have likely resulted in such massive attrition to fighting forces as to make reinforcement and resupply questionable (although not irrelevant). Major ports and unloading facilities would have quickly become unusable. Had the Soviets been the first to resort to nuclear weapons, perhaps in a pre-emptive strike against NATO nuclear weapons and airfields, allied maritime forces would, however, have an important role in a retaliatory strike. Whatever the uncertainties and doubts about the ability of NATO to conduct a selective nuclear campaign in Europe, it remained the case that the Alliance had a fairly secure theater nuclear capability at sea. Carriers could deliver nuclear strikes from ranges beyond those optimal for close support of a conventional land battle. The deployment of cruise missiles on American surface and subsurface units afforded NATO an additional theater capability (see Zimmerman and Greb 1982; Sorrels 1983: 72). The United States had, for nearly 20 years, designated a portion of its SSBN fleet available to NATO. With continual MIRVing and greater accuracies, the SLBMs on these submarines could be used in a theater capacity.

The use of nuclear weapons in the land and air war would not automatically have made efforts to secure the sea lines of communication moot. Much would have depended on the nature of a European nuclear exchange. For example, the Soviets might have opened an attack with a limited series of nuclear strikes, intended to pave the way for conventional forces and paralyze the NATO partners' ability to take collective action. Whether NATO responded with a nuclear strike of its own, or absorbed a limited blow hoping either to keep the war conventional or negotiate, the sea lines of communication from North America and England, as well as those in the Mediterranean, would have to have been secure. Even if Soviet strikes at major ports prevented the large-scale reinforcement of the land and air forces, some military sealift movement would have had to have continued until NATO decided either to seek a settlement or escalate. And, even in the context of a post-exchange ceasefire, there would have been a requirement for civilian shipping, necessitating continual surveillance of remaining Soviet maritime forces. As long as NATO maintained a position on the continent, either through early resort to negotiation or sustained conventional resistance in a post-exchange setting, allied maritime forces would have had to continue efforts to secure the SLOC. Not to do so would have simply added one more element of peril to the land and air situation.

In the event of a collapse on the continent, military or political, victory in the war at sea, even a nuclear one, could not have conceivably held open the prospect of a reversal of a Warsaw Pact conquest. The United States, and perhaps Britain if it had survived unscathed, would have maintained forces at sea. This would include SSBN forces which would still constitute a deterrent. If strategic nuclear war followed or preceded the final collapse of allied forces in Europe, these sea-based nuclear forces would undoubtedly have been both used and subject to attack, as would those of the Soviets.

CONCLUSION

The recent debates and controversy over the maritime strategy have tended to confuse and obscure an understanding of the role which maritime forces had always played in NATO's nuclear, and especially conventional, deterrent. As part of its aggregated forces, the Alliance looked to its contributing navies, principally the USN, to secure, deny and exploit the seas in support of the allied position on the European mainland.

During the early days of the Cold War, NATO enjoyed an overwhelming margin of maritime superiority. Had a conflict broken out in Europe, the USN would quickly have been able to secure the sea line communication across the Atlantic and to press the war at sea forward, projecting conventional and nuclear power ashore, including against the USSR itself. In the 1950s, the Alliance was, however, counting on American strategic nuclear superiority as its principal deterrent. Thus its maritime dominance was not considered crucial to the overall posture. NATO could not safely ignore its maritime forces and indeed did not by any means, but the development of those forces did not rank high on the allied list of priorities.

As the strategic balance shifted from American superiority toward a rough equivalence, and as Warsaw Pact conventional forces continued to grow and maintain their superiority, NATO began to move, however unevenly and ambiguously, toward a posture which emphasized deterrence below the strategic nuclear level. Flexible response declared that the Alliance would respond to an attack at the level chosen by the aggressor. It did not rule out the first use of nuclear weapons to prevent an allied conventional defeat, but it did postulate a greater reliance upon conventional forces. The allies began to consider measures that would allow it to sustain a conventional forward defense along their entire perimeter, but especially the central European front, for several weeks.

Flexible response increased the importance of the American and other allied naval contributions to NATO well before the USN's articulation of its 'maritime strategy'. While there was not then, nor has there ever been, any certainty that conventional defense would in fact be possible, or even that a war in Europe would begin or long stay at that level, flexible response increased the demands placed upon allied maritime forces. The basic tasks remained the same – to secure the seas for conveyance and projection – but now support of the land and air forces ashore had assumed more urgency and importance. Not only would sealift be essential for sustained conventional combat, the ability of the Alliance to support its position ashore through seapower became an additional hedge against the need to resort to the first use of nuclear weapons.

As NATO was placing more reliance upon its maritime forces to perform its traditional tasks, the capacity of the Warsaw Pact to deny NATO use of the seas had increased markedly. Quantitatively and qualitatively, the Alliance still held a margin of superiority, but it was a diminishing margin. Particularly in the European coastal waters, NATO faced the prospect of a prolonged struggle at sea before it could safely begin to sealift forces and project force ashore. For the Alliance, the combination of the adoption of the strategy of flexible response and the capacity of the Soviets to deny NATO free use of the seas highlighted the importance of conventional maritime forces. In spatial terms, it meant that NATO had to maintain a wide range of maritime forces from strike carriers to minesweepers; and, temporarily, it meant that the Alliance had to be more concerned with pre-hostilities sealift, the forward deployment of maritime forces and the capacity to sustain land and air forces in a conventional war of indeterminate length.

The decade of the 1970s saw NATO give greater attention to its maritime posture. A standing naval force was created under the Atlantic Command. While not a significant fighting force, STANAVFORLANT provided valuable training and practice in allied naval co-operation. In the Mediterranean, there was a re-organization of the command structure and the creation of a new subordinate command for maritime–air co-operation. Procedures for the timely acquisition of NATO flag shipping for the purpose of sealifting American reinforcements and resupply were completely reworked and updated. Despite budgetary pressures, the USN and other NATO navies continued to modernize. At the end of the decade, the three major military commanders had finalized an accepted 'concept of maritime operations'.

While NATO had become more of a maritime alliance in the 1970s, the importance of seapower in the overall European balance of power continued to be circumscribed by the existence of nuclear weapons. Without nuclear weapons, the deterrent value of the Alliance's still considerable seapower would have been far greater, because in a protracted struggle NATO could still have secured use of the sea allowing for massive reinforcement and resupply. However, in the presence of a strong likelihood that nuclear weapons would be used within a matter of weeks, if not days, in the event of a conflict in central Europe, the relative importance of the USN's and other allied navies' conventional capabilities remained inherently limited.

Flexible response by no means ruled out the use of nuclear weapons. NATO still relied upon the threat of first use as its ultimate deterrent and the USN's maritime forces continued to constitute an important element in the Alliance's nuclear posture at the theater/tactical and strategic levels. But, given the infinite imponderables surrounding use of these weapons, even if that use had been restricted to the war at sea, the ability of maritime forces to perform their tasks would have carried the same uncertainties as land-based atomic forces.

Apart from the nuclear factor, the relative importance of NATO's maritime forces was circumscribed by the continuing imbalance of land and air forces, particularly along the central front. The effectiveness of seapower remained linked to the situation ashore. The ability of the USN and other navies to exploit the seas such as to impact favorably upon a conventional land and air war still depended upon the outcome of the early stages of such a conflict. Sustained conventional resistance would have required massive sealift, but even with a pre-hostilities mobilization of shipping, large-scale support would have had to have awaited the outcome of the initial battles at sea. In the meantime, the various land and air fronts would have had to have held off superior Warsaw Pact forces able to use internal land lines of reinforcement and resupply. Thus, an eventual 'victory at sea' in a conventional conflict would have been far from decisive for NATO. The Soviets could lose the war at sea and still have pushed deeply into allied territory, confronting the Alliance with the choice between surrender or escalation. And Moscow was now able to rely upon its own considerable theater/tactical and strategic nuclear forces to deter the NATO governments from exercising their first use option.

By the late 1970s, allied maritime forces, while indispensable to the strategy of flexible response, provided no solutions to the unavoidably ambiguous and uncertain character of that strategy. The condition of mutual nuclear deterrence and the continuing imbalance of conventional forces both compelled NATO to place greater emphasis upon its maritime forces while simultaneously circumscribing the relative importance of the

Alliance's collective seapower, including that of the United States. With CONMAROPS and its own 'maritime strategy', the USN sought to compensate for these limitations by devising better means to exploit its still superior position at sea, thus providing for better protection of the sea lines of communication and more effective projection of force ashore. Thus, most of the 'maritime strategy' of the 1980s was by no means inconsistent with the importance of the United States Navy to NATO since 1949. Critics of the strategy tended to ignore the role American seapower had always played in European security, especially since the advent of flexible response. It was one of the bonds that helped hold the transatlantic bargain together.

At the same time, given the pessimistic assessments offered by the United States Navy itself in the late 1970s concerning its ability to contribute to the defense of Europe in the event of a NATO/Warsaw Pact war, the claims made by the USN for its 'maritime strategy' were somewhat surprising. Like flexible response and CONMAROPS, the supposedly new concept of the role of seapower in the Alliance's posture seemed to assume a measure of maritime superiority which no longer existed. With predictions that NATO would have a difficult time securing use of the ocean approaches to Western Europe and the Norwegian Sea, it was difficult to accept the USN's calls for operations inside Soviet SSBN bastions and global horizontal escalation of the war at sea. To this extent, some of the criticism of the maritime strategy as an unrealistic, indeed almost romantic, return to Mahanian thinking was well justified. The Alliance was similar to the allied coalitions of the past, which had been able to counter a predominant land power by avoiding excessive reliance upon fleet-against-fleet capabilities, and concentrating on using seapower to support and sustain forces ashore.

Precisely because American naval power had been so important in sustaining the cohesion and unity of the Alliance, the unilateral tendency of the 'maritime strategy' was also a matter of legitimate concern. Thus critics were equally right in stressing that a 'maritime strategy' was no alternative to flexible response as a framework for the military and political decisions that guided the Alliance and indeed had helped keep the allies allied. The difficulties posed by flexible response to NATO's naval posture simply affirmed at the end of the Cold War what had been true at the beginning – that the effectiveness and significance of seapower in war and peace is integrally linked to the broader strategic and political environment.

Notes

1 Introduction

1 For the official USN explanation of the strategy see Watkins (1986). Also contained in this supplement are statements by General P. X. Kelly, Commandant of the Marine Corps, and John F. Lehman, Secretary of the Navy and a major proponent of this strategy. See also Troost (1987); Palmer (1988); Friedman (1988); Gray (1986).
2 For the flavor of this aspect of the debate see Vlahos (1982); Dunn and Staudenmaier (1984); Mearsheimer (1986); Brooks (1986); Komer (1984); Barnett (1987).
3 The view that the Alliance's maritime posture was ignored can be found in such works as Myers (1979).
4 See, for example, Nitze *et al.* (1979); Bertram and Holst (eds) (1977); George (1978); Wilson (1976); Kidd (1978); Moulton (1974); Cottrel and Moorer (1979); Wegner (1975); Till *et al.* (1982).
5 For the purposes of this study, theater and tactical nuclear weapons have been placed together as forming the second element in the NATO triad. Tactical or battlefield nuclear weapons (or short-range theater nuclear weapons) are those 'designed to influence directly the outcome of combat by destroying engaged enemy forces'. Their warhead yields range from 0.1 to 2.0 kilotons and can be fired distances of less than 70 miles. Examples of tactical nuclear weapons would be artillery shells, Lance and Honest John missiles.

Theater nuclear weapons are weapons 'designed to influence indirectly the outcome of combat by interdicting the movement of enemy troops and supplies to and from the battle area or by destroying rear area installations – airfields, marshaling yards, supply depots – vital to the enemy's continued prosecution of hostilities'. Included as well would be destruction of enemy theater nuclear forces. These weapons have a range of 400 miles and yields of 3 to 400 kilotons. The Pershing I missile would fall into this category.

Another class of weapons systems has been labeled 'semi-strategic nuclear weapons'. This would include weapons such as the F-111 and forward-based fighter bombers that are 'designated by the United States for theater use but could reach and return from targets in the Soviet Union'. Carrier-based nuclear-capable aircraft would also fall into this category.

Weapons designed for combat at sea would fall into the theater category. These would include ASW depth charges and missiles, i. e. SUBROC, and nuclear armed ASMs and SSMs.

The above definitions were drawn from Record (1974: 6–7).

6 Generally excluded from the theater class of nuclear weapons are long-range strategic bombers, ICBMs and SLBMs. However, several American SSBNs and the entire British SSBN force are earmarked for SACEUR. In 1975, the US assigned several SSBNs carrying the new Poseidon MIRV (Multiple independently targeted re-entry vehicle warhead) SLBM to NATO. The Poseidon is said to be capable of 'selective, discriminatory attacks on targets close to the battlefront, designed to stop the Warsaw Pact forces'. SACEUR incorporates available SLBMs in its GSP (General Nuclear Strike Plan) in co-ordination with the US Single Integrated Operational Plan (SIOP), which governs the use of American strategic nuclear forces. SIPRI (1978: 115); Finney (1975); US Senate (1973); Schlesinger (1977).

2 Establishing the NATO Maritime Alliance

1 US Navy Operational Archives (OA), Papers of Admiral Sherman (SP), Box 9 File: 'Presentation to President, House and Senate Committees'.
2 OA, A16–3(5) War Plans 1947, OP 30 File, Fleet Admiral C. W. Nimitz, Chief of Naval Operations to Commanders-in-Chief Atlantic, Pacific and Naval Forces Eastern Atlantic and Mediterranean, 'Tentative Assignment of Forces for Emergency Operations', 12 July.
3 Submission of the Commander-in-Chief Atlantic Fleet to the General Board Study, 26 April 1948: 3, Item 104. (In completing its study, the General Board sent out a questionnaire containing over a hundred items, or topics, upon which naval commanders and numerous other individuals were to comment. The General Board final study was based largely on the responses to this questionnaire. As such it represents the best summary of naval thought during these years.)
4 OA, Strategic Plans, A16–1, Miscellaneous Plans and Studies, Box 4, Envelope 105, 'Navy Strategic Plans 1950'.
5 Ibid., Box 3, Envelope 65, 'Airlant Staff Study on ASW Plans'.
6 OA Airlant Study, op. cit.
7 OA, Strategic Plans, A16–1, Miscellaneous Plans and Studies, Box 2, Envelope 53, 'CNO Annual Report Folder'.
8 OA, Immediate Files of the CNO (OP–00), 1949, Box 8, 'North Atlantic Ocean Regional Planning Group, First Meeting', 31 October.
9 OA Command Files (CF) 1952, Post 1 January 1946 United States Department of the Navy, CINCNELM (Commander-in-Chief Eastern Atlantic and Mediterranean), 'Annual Report of Operations and Conditions of Command During Fiscal Year 1952', July 1951–14 June 1952: 2.
10 OA (SP), Box 6, File CNO European Trip March–2 April 1952, Serial 000158P40, 20 March 1950, 'Proposed Trans Atlantic Trip of Chief of Naval Operations', Enclosure 7, 'Statement of Fiscal 1951 MDAP Program'.
11 US National Archives, 1951, Records of the Joint Chiefs of Staff (RJ), JCS 902 Western Europe 3–12–48 (WE), JCS 2073/153, 'Division of Command

Responsibility between SACEUR and SACLANT in the North Sea Area', 16 May.
12. Library of Congress (LC) 1952, McCormick Papers (MP), Box 6, File 7, 'SACLANT Initial Trip to Atlantic Command Nations, 24 February–20 March 1952, Informal Report to Standing Group, 1 April'.
13. MP, Informal Report, op. cit.
14. Ibid.
15. MP 1952, Box 6, File 7, 'Remarks Before the Permanent National Representatives to the North Atlantic Council', 28 May 1952.
16. Ibid.
17. OA, 1952, Papers of the Chief of Naval Operations, Strategic Plans, Box 1, File A16–22, 'From: Director General Planning Group To: Deputy CNO, NATO Naval Forces for D-Day 1953'.
18. Ibid.
19. OA (CF) 1953, US Navy, Annual Report of the Commander-in-Chief Atlantic Fleet, 1 July 1952–30 June 1953, CINCLANTFLT Report: 1.
20. Ibid.: 21
21. Ibid.: 28
22. Formally, SACLANT had requested that the United States Liaison Officer to SACLANT obtain from CINCLANT the information on the special convoys planned by the JCS. Since McCormick was both SACLANT and CINCLANT, it would appear that the request was essentially one for permission to share the information with other NATO naval staff officers and commanders.
23. RJ CCS 902 (3–12–48), Sec. 167, 'From: Commander-in-Chief Atlantic To: US Liaison Officer, SACLANT, Subj: SACLANT Plans for Special Convoys in North Atlantic Area', 25 August.
24. Ibid.
25. RJ (WE) 1952, Joint Strategic Plans Committee (JSPC) 876/573, 'Report to the Joint Chiefs of Staff on SACLANT Emergency Defense Plan', 29 October.
26. OA (CF) 1954, US Navy, Annual Report of the Commander-in-Chief Atlantic Fleet (Supplementary), 12 April – 30 June 1954, CINCLANTFLT Supp.: 14; interview with Admiral Jerauld Wright, SACLANT 1954–60, 23 March 1981.
27. RJ (WE) 1952, JCS 2073/478, 'Memorandum by Chief of Naval Operations to Joint Chiefs of Staff on SACLANT Emergency Defense Plan', 28 November.
28. OA (SP) 1951, Box 7, SHAPE Folder, Telegram, SACEUR to CNO, 15 March.
29. OA (SP) 1951, Box 7, Mediterranean Folder, Telegram, CNO to SACEUR, 20 May.
30. OA (SP) 1951, Box 7, Mediterranean Folder, Memo to Sherman, Serial 0000607P30, 'Deployment of US Carriers to Mediterranean and General European Area', 2 June.
31. RJ (WE) 1952, JCS 2073/473, 'Memorandum by the Chief of Staff, US Air Force, for the Joint Chiefs of Staff on SACLANT Emergency Defense Plan', 24 November.
32. Ibid.
33. RJ (WE), JCS 2073/478, op. cit.
34. Ibid.

35 MP 1953, Box 6B, File on Mariner Trip, 'SACLANT Memorandum to Admiral Carney, USN, Chief of Naval Operations, Subj: Exercise MARINER, 16 September to 4 October 1953', 13 October.
36 Ibid.
37 RJ (WE) 1952, JCS 2073/436, 'CNO Memorandum: Assignment of Naval Forces to SACLANT', 8 October.
38 OA (CF) 1954, US Navy, Annual Report of the Commander-in-Chief Atlantic Fleet, upon being relieved, 1 July 1953, 12 April 1954, CINCLANTFLT Report: 52.
39 OA (CF) 1955, US Navy, Annual Report of the Commander-in-Chief Atlantic Fleet, 1 July 1954 – 30 June 1955, CINCLANTFLT Report: 61.
40 Ibid: 18.
41 OA (CF) 1954, CINCLANTFLT Supp., op. cit.: 14.
42 OA (CF) 1955, CINCNELM, 'Report of Operations and Conditions of Command During Fiscal Year 1955, 1 July 1954–30 June 1955', CINCNELM: 28.
43 This account of the early command arrangements and strategic considerations involved in NATO maritime strategy in the Mediterranean is drawn mostly from Mountbatten (1955).
44 Ibid.: 173.
45 Based on an interview with Admiral I. Kidd, USN (Ret.), former Commander of the Sixth Fleet, at Falls Church, Virginia, 23 March 1982.
46 'The Baltic Approaches', *NATO Letter*, Vol. 13, No. 1, January 1965: 23.
47 Ibid.
48 In their futuristic account of a third world war, General Sir John Hackett and associates have NATO creating a new supreme naval command, Joint Allied Command Western Approaches (JACWA). The new command incorporates part of the EASTLANT within the UK Air Defence Region and all of CINCHAN. According to the authors, various changes in the military situation, including technological advances in weaponry and communications, the increase in the number of Soviet submarines and 'most important' the adoption of a strategy of 'flexible response' meant that 'while the basic requirement remained, that the NATO ground forces should be reinforced and supplied across the Atlantic, almost every other condition which governed the original command structure had altered'. Flexible response required that the Alliance maintain a credible peacetime presence at sea to engage the enemy from the outset. The speed with which conventional war would develop meant that the whole of north-central Europe, including the sea areas, would have to be regarded as part of the battlefield from the outset as well as being the 'terminal and entrepot for transatlantic support'. Under these circumstances, 'the purely political character of Channel Command, squeezed between a subordinate command of SACLANT, who was far away in Norfolk and the presumably hard-pressed SACEUR, across the Channel, could not be expected to meet the requirements of modern war'.

JACWA was to be a joint air–sea command under the British CinC for Air for the Fleet, with the Royal Navy's submarine commander and an American flag officer as deputies, and an Admiral of some other nationality as Chief of Staff. The new command would seek to keep the sea lanes open and, early on in a war, be engaged in a battle for the Baltic exits. (Hackett *et al.* 1978: 343–5)

3 The Cold War at sea: force and strategies

1 This overview of the European navies in the early 1960s draws upon AWEU (1963) and Kennet (1964a).
2 Speech by Admiral Sola, French Navy, as quoted in *NATO Letter*, Vol. 11, No. 11, November 1963: 28.
3 OA (CF) 1967, US Navy, Annual Report of the Commander-in-Chief Atlantic Fleet, 1 July 1966–17 June 1967, CINCLANTFLT Report: 111.
4 OA (CF) 1959, US Navy, Commander-in-Chief, US Naval Forces Eastern Atlantic and Mediterranean, Annual Historical Report, 1 January 1959–31 December 1959, CINCNELM Report: 4.
5 US Naval War College, Newport, Rhode Island, Naval Historical Collection, The Richard G. Colbert Papers (CP) Series 1, Box 14, Folder 285, Letter from Admiral G. Pighini, Commander Allied Naval Forces South to Admiral Richard Colbert, CINCSOUTH, 12 September 1973.
6 United Kingdom, Central Office of Information, Fact Sheet, Defence (No. 17058/FS/81).
7 In 1968 Canada unified its armed forces. Abolished were the Royal Canadian Air Force (RCAF), Royal Canadian Navy (RCN) and the Canadian Army. The unified forces was divided along 'command' function lines and, after further re-ordering, Air Command, Maritime Command and Mobile Command took the places of the old RCAF, RCN and the Army. The Commander of the Maritime Command assumed NATO duties as subordinate Commander of the CANLANT region under SACLANT. At his disposal are Canadian ships stationed in the Atlantic and naval aviation units drawn from Air Command and placed under MARCOM's operational command.
8 MP, Box 6, File 7, 'SACLANT Trip Report'.
9 OA (CF) 1963, US Navy, Annual Report of the Commander-in-Chief Atlantic Fleet, 1 July 1962–30 April 1963, CINCLANTFLT Report: 6.
10 OA (CF) 1958, US Navy, Annual Report of the Commander-in-Chief Atlantic Fleet, 1 July 1957–30 June 1958, CINCLANTFLT Report: 66.
11 OA (CF) 1959, CINCNELM Report, op. cit.: 11–12.
12 OA (CF) 1964, US Navy, Annual Report of the Commander-in-Chief Atlantic Fleet, 1 May 1963–30 June 1964, CINCLANTFLT Report: 64.
13 OA (CF) 1967, CINCLANTFLT Report op. cit.: 111.
14 OA (CF) 1959, US Navy, Report of the Commander-in-Chief Atlantic Fleet for the Period 1 July 1958–30 June 1959, CINCLANTFLT Report: 41, 73.
15 OA (CF) 1960, US Navy, Report of the Commander-in-Chief Atlantic Fleet for the Period 1 March 1960–30 June 1960, CINCLANTFLT Report, June: 37.
16 OA (CF) 1957, US Navy, Annual Report of the Commander-in-Chief Atlantic Fleet, 1 July 1956–30 June 1957, CINCLANTFLT Report: 29–35; OA (CF) 1958, Annual Report of the Commander-in-Chief Atlantic Fleet, 1 July 1957–30 June 1958, CINCLANTFLT Report: 33.
17 CINCNELM Report, 1959, op. cit.: 18–19, 22.
18 OA (CF) 1961, US Navy, Report of Operations and Conditions of Command, Commander-in-Chief, US Naval Forces, Europe, 1 July 1960–30 June 1961, CINCUSNAVEUR Report: I–18.
19 CINCNELM Report, 1959, op. cit.: 26.

20 OA, Short of War Documentation (SWD) 1963, Special List No. 3, 'Keeping the Peace Chronology', 22 August: 27.
21 OA (SWD) 1977, Robert B. Mahoney, *U.S. Navy Responses to International Incidents and Crises, 1955–1975*, Vol. II, Arlington, Virginia, Center for Naval Analysis, July: C–22.
22 OA (CF) 1963, US Navy, Commander-in-Chief US Naval Forces Europe, Report on Operations and Conditions of Command, 10 April 1963–26 June 1963, CINCUSNAVEUR Report, June: 1–2.
23 OA (CF) CINCLANTFLT Report, 1962, op. cit.: 5.
24 OA (CF) 1965, US Navy, Annual Report of the Commander-in-Chief Atlantic Fleet, 1 July 1964–30 April 1965, CINCLANTFLT Report: 25; CINCLANTFLT Report, 1967, op. cit.: 45.
25 CINCUSNAVEUR Report, June 1963, op. cit.: 1–2.
26 OA (CF) 1964, US Navy, Commander-in-Chief US Naval Forces Europe, Report on Operations and Conditions of Command, 26 June 1963–30 June 1964, CINCUSNAVEUR Report: 1–3.
27 Interviews: Geoffrey Till, Royal Naval College, Greenwich, UK, 21 March 1983; Ministry of Defence, London, 23 March 1983.
28 There was one tangible outcome to the MLF efforts. In 1962 the Italian Navy commissioned the rebuilt cruiser *Garibaldi* as a guided missile ship armed with Terrier anti-aircraft missiles but also equipped to fire a Polaris-type missile.
29 US Dept of Defense 1950 (Contract No. N5 ori 07846). The study was done at the request of the United States Chief of Naval Operations, Vol. I: A–19.
30 OA (CF) 1956, US Navy, Annual Report of the Commander-in-Chief Atlantic Fleet, 1 July 1955–30 June 1956, CINCLANTFLT Report: 31.
31 OA (CF) 1962, US Navy, Annual Report of the Commander-in-Chief Atlantic Fleet, 1 July 1961–30 June 1962, CINCLANTFLT Report: 5.
32 For a historical account of American nuclear ASW weapons see Sokolsky (1984).
33 OA (SWD), Berlin File. Burke also requested a legal opinion on whether harassment of Soviet shipping would be justified under international law as a response to Soviet harassment in Berlin. The study concluded that such measures would be legal.
34 US Navy, *War at Sea*, Vol. I: 2. The full study remains classified; quotations and references made here are from a 'sanitized' extract made available by the Office of the Chief of Naval Operations.
35 Ibid., Appendix IV: IV–10.
36 Ibid.: 10.
37 Interviews: Geoffrey Till, op. cit., Admiral of the Fleet, Lord Hill-Norton, London, 22 March 1983.
38 CINCLANTFLT Report, 1962, op. cit.: 79.
39 OA (SWD) 1966, *Bendix Report: The Navy and Sub-limited Conflicts*, 20 September: A–65.
40 CINCLANTFLT Report, 1967, op. cit.: 12, 150.
41 OA (CF) 1958, US Navy, Annual Report of the Chief of Naval Operations to the Secretary of the Navy For Fiscal Year 1958: 7.
42 See, for example, Martin (1965); Martin and Bull (1968); Guillon (1968); Gretton (1965); Dillon (1972).

43 The most well known of the bureaucratic-orientated analyses of post-Second World War naval strategic development were: Davis (1966 and 1967); Hammond (1963); Stein (1963).
  See also Dillon (1972). Dillon argues on the basis of Martin's analysis that Canadian naval construction which stressed ASW escorts was essentially a departure from what normative decision making would have dictated. The distortion was brought about by Canada's naval leaders failing to take into account the implications of nuclear weapons on naval roles. A rational approach would have identified the misconception of believing that in a future war convoy escorts would be needed. Whether this was an irrational judgement remains in question. Martin did not say all ASW preparations were useless; he stressed that the West would have to maintain a limited capability if only to keep up with the technology. Moreover, it is hard to see Canada, a relatively small maritime power, not following the overall decisions regarding naval posture undertaken by the Alliance, and by the United States. So long as Canada remained part of NATO, its maritime contributions, as with its other standing forces, would largely be based upon the strategic posture adopted by its allies.
44 As quoted in *NATO Letter*, Vol. 5, No. 2, February 1957: 27–30; ibid., Vol. 5, No. 12, December 1957: 26–8.
45 RJ (WE) 1951, JCS 1868/258, Memo From General Bradley to Secretary of Defense, 24 April.
46 OA (CF) 1957, CINCLANTFLT Report, op. cit.: 27.
47 OA (SWD) 1963, Keeping the Peace Chronology, op. cit: 27.
48 OA (CF) 1962, US Navy, Annual Report of the Commander-in-Chief Atlantic Fleet, 1 July 1961–30 June 1962, CINCLANTFLT Report: 9.
49 OA (CF) 1955, CINCLANTFLT Report: 14.
50 NATO (1969). This was the *Brosio Study*, carried out by SACLANT at the request of the then Secretary General Manlio Brosio. Further information on the findings of the study are covered in the next chapter. The *Brosio Study* can be found at SACLANT Headquarters in Norfolk, Virginia. Much of it remains classified. The author was allowed to read the study in its entirety on the understanding that only general, summary references would be made as to its contents. Thus references to it, as in this case, and in the next chapter, contain no direct quotations nor quantitative material.

4 Soviet maritime forces and flexible response

1 US Congress (1973: 104). Statement by General Goodpaster.
2 US Congress, *Congressional Record*, 12 June 1972, Washington DC: GPO, 1972: 20493. In 1972, Senator Proxmire and Chief of Naval Operations Elmo Zumwalt engaged in an exchange of letters regarding the US–Soviet naval balance. The Senator claimed that the US Navy was still far more powerful than the Soviet fleet measured in terms of the quantity and quality of forces. Zumwalt stressed that the missions of the two navies were different, with the US Navy having a protective and projective function and the Soviets a sea denial, and therefore the US needed better forces. Hereafter cited as the Proxmire–Zumwalt Exchange.

3 By the early 1970s a number of the European NATO navies had begun developing SSM. These included the Penguin deployed on Oslo Class Norwegian destroyers and small patrol boats, and Exocet missiles deployed on French, German and Greek ships (see *Jane's* 1974).
4 Proxmire–Zumwalt Exchange, op. cit.: 20501.
5 See, for example, the special issue of *NATO's Fifteen Nations*, Vol. 17, No. 1, March 1972, with articles by NATO's major naval commanders; Wegner (1975); Zumwalt (1976).

The best available documentary sources are the papers of Admiral Richard G. Colbert, USN (CP) held at the US Naval War College Historical Collection, Newport, RI. Colbert served on the staff of SACLANT and was CINCSOUTH at the time of his death in 1973. He also kept a continuing correspondence with many allied naval officers. Two items of particular interest are CP, Series I, Box 19, Folder 369, the exchange of letters with Commodore Kierkegaard, Royal Swedish Navy, and CP, Series II, Box 18, Folder 38, the talking notes to the presentation 'Soviet World Wide Threat'. References to the Colbert Papers (CP) use the list found in the Register listing the holdings. Because many items have been removed for security reasons the citations given below may not correspond to the exact location of the documents.
6 In addition to items cited above, see also Blechman (1973); Polmar (1972); Breyer (1970); Kilmarx (1975); US Senate (1976); Watson (1982).
7 US Senate 1976: 157. Testimony of Michael MccGwire.
8 This possibility was suggested to the author by officers of Allied Command Channel at a briefing, 24 March 1983.
9 US Congressional Budget Office, 1976, Washington DC: CBO, 26 March: 29.
10 Briefing at Channel Command, 24 March 1983; interview with former Admiral of the Fleet, Lord Hill-Norton, 22 March 1983.
11 As late as 1976, the combined land-based air forces of Italy, Greece and Turkey included just over 400 capable aircraft. Facing NATO in the south were nearly 5,000 Warsaw Pact tactical aircraft belonging to the WP air forces. Soviet naval aviation disposed of another 650 tactical aircraft, 700 long-range naval aircraft of which 200 were Bear, Bison and Backfire bombers.

The two Sixth Fleet carriers brought 48 fighters, 46 attack aircraft, 24 all-weather planes and 8 early warning planes to NATO. These were the most modern F-14, A-7 and A-6 aircraft in the NATO inventory. An additional carrier was earmarked for the Mediterranean during war, thus increasing by 50 per cent the contribution of carrier-based airpower (US Congress, Senate 1977: 3–4).
12 Letter from Admiral Rivero to author, 16 May 1983; see also Dur (1976: 105); Rivero (1978: 524).
13 CP (Series I, Box 15, Folder 347), Letter from Admiral G. E. Miller, Commander US Second Fleet, NATO Striking Fleet Atlantic, to Rear Admiral Stansfield Turner, undated.
14 'Exercise Strong Express in Retrospect', *International Defense Review*, No. 6, 1972.
15 CP, op. cit., letter from Miller to Turner.
16 'Exercise Strong Express in Retrospect', op. cit.: 662.

17 OA (CF) 1967, US Navy, Annual Report of the Commander-in-Chief Atlantic Fleet, 1 July 1966–17 June 1967: 150.
18 CP (Series II, Box 18, Folder 38), 'The Soviet World Wide Maritime Challenge', a briefing prepared by headquarters, Supreme Allied Command, Atlantic.
19 CP (Series I, Box 19, Folder 369), Letter from Richard Colbert to C. Kierkegaard, 27 June 1968.
20 Ibid.
21 Ibid.; interview with Admiral G. E. Miller, USN (Ret.), 22 April 1983.
22 CP (Series I, Box 17, Folder 330), Letter from Richard Colbert to Captain J. A. Merin, 17 February 1967; letter to C. J. Zimmerman, 1 March 1967.
23 CP (Series I, Box 18, Folder 331), Letter to W. Erikson, USN, 4 May 1967.
24 CP (Series I, Box 18, Folder 331), Letter to Walter Rostow, 4 May 1967.
25 CP (Series I, Box 19, Folder 335), Letter to Captain Gillow, FN, 1 November 1967.
26 CP (Series I, Box 19, Folder 366), Letter to Rear Admiral J. H. Adams, RN, 20 December 1967.
27 CP (Series I, Box 18, Folder 337). The Colbert Papers contain two declassified briefings on STANAVFORLANT, 'Naval War College Look at Standing Naval Force Atlantic' enclosure to a letter to RADM, L. Gies, 18 February 1969; CP (Series II, Box 21, Folder 39) 'SNFL Briefing for CINCL-ANT/CINCLANTFLT Officers' (undated).
28 CP, SNFL Briefing, op. cit.
29 Ibid.
30 Interview with Former Admiral of the Fleet, Lord Hill-Norton, 22 March 1983; interview with senior officers United Kingdom Ministry of Defence, 23 March 1983.
31 SNFL Briefing, op. cit.
32 CP (Series I, Box 18, Folder 350), Colbert outlined a proposal for the use of the *Savannah* in 'N. S. *Savannah*: A Concept for Use as Flagship for Standing Naval Force Atlantic', enclosure to letter to Zumwalt, 23 October 1971.
33 CP (Series I, Box 18, Folder 350), Letter from Zumwalt to Colbert, 17 November 1971.
34 Duncan (1972: 35); interview, Admiral J. H. Scheuer, 15 March 1983.
35 Rivero (1972: 3); letter from Admiral Rivero to the author, 16 May 1983.
36 Interview with Admiral Gerald E. Miller, USN (Ret.), 22 April 1983.
37 Ibid.
38 'Fourth Activation of NATO's Naval On-Call Force Mediterranean', *NATO Review*, Vol. 20, Nos 7–8, July–August 1972: 24.
39 See especially CP (Series II, Box 18, Folder 38) for an account of the steps leading up to the initiation of the *Brosio Study*. As indicated in Chapter 3, the *Brosio Study* can be found at SACLANT headquarters, Norfolk, VA. References to its findings here are only summaries of the still-classified portions.
40 CP (Series II, Box 21, Folder 45), Letter from Zumwalt, 6 September 1970. The *Newport Study* itself remains classified.
41 CP (Series I, Box 18, Folder 348), Letter to Zumwalt, 3 September 1971.
42 US Senate (1978: 1737). Testimony by Admiral Crowe.
43 Mr Maurice Foley, Under Secretary of State for the Navy, as quoted in Grove (1987).

44 In 1981, the British government decided that it would retire nine of its 59 frigates and the ASW carrier *Hermes* and stress the combination of ASW aircraft and attack submarines in meeting the Soviet submarine threat. Surface ships were regarded as too vulnerable and less cost-effective ASW platforms. While this could leave a gap in surface forces, the improvements in air and submarine ASW reaffirmed the UK's dedication to the allied maritime posture (see Chichester and Wilkinson 1982: 133).
45 Briefing at CINCHAN, 24 March 1983.
46 US Senate 1978: 1616. Testimony by Ambassador Robert Komer.
47 'Exercise Strong Express in Retrospect', op. cit.; Marriott (1973); US Dept of Defense (1972).
48 *NATO Review*, July–August, 1972:1.
49 US Senate 1978: 1656. Testimony by Ambassador Robert Komer.

5 Reinforcement sealift

1 See, for example, the testimony of Dr Robert Kilmarx, Director of Business and Defense Studies, Center for Strategic and International Studies in US Congress (1980a: 65–73).
2 US Dept of Defense 1974, *Annual Report FY 1975 and FY 1975–1979 Defense Program*, Washington DC: GPO: 166.
3 RJ (WE) 1952, JCS 3-12-48, Joint Military Transportation Committee (JMTC), JMT 166/65 'Control of the Use of Shipping', 31 March.
4 A central logistical planning problem is to determine how many ships, or other methods of transportation, would be required to transport a given quantity of cargo from one point to another according to a particular schedule. It is, however, extremely difficult to relate the capacity, speed and cargo-handling characteristics of each specific carrier to the demands of a lift requirement schedule. Planners, in the case of ships, have adopted the concept of a 'notional ship'. A notional ship is not any particular vessel, but rather represents a mythical ship with given characteristics, such as carrying capacity, speed and sometimes turnaround time in port. It is a standard of measurement of sealift capability. The characteristics behind the notional ship vary from assessment to assessment. In the above-quoted document no indication is given of the characteristics of the notional ships.

As an example, if one used a notional ship as a ship able to carry 15,000 metric tons, at a given speed and turnaround time such that it could make a transatlantic crossing and unload and return in 46 days, then to move 30,000 metric tons to Europe (returning empty) in 46 days would require 2 notional ships. An actual ship with similar speed but with 30,000 metric ton capacity would be said to have the value of 2 notional ships. So that the notional value of a particular ship can be expressed as follows:

$$NV_a = \frac{C_a}{C_n} \times \frac{(TA)_n}{(TA)^n}$$

where a denotes an actual ship
n denotes a notional ship   c = capacity
(TA) = ship turnaround time in where days which is a function of speed where and unloading capability
NV = notional value

(Hunt and Rosholdt [n.d.])

5 RJ (WE) 1952, Sec. 120, JCS 2073/300, 'SHAPE's Shipping Requirements', 8 February: 1964.
6 Ibid.: 1965.
7 Ibid.: 1968.
8 Ibid.: 1968-9.
9 Ibid.: 1969.
10 Ibid.: 1972.
11 Ibid.: 1975.
12 RJ (WE) JMTC 166/65, op. cit.
13 Ibid.
14 US Maritime Administration 1978, *Annual Report Fiscal 1977*, Washington DC: GPO: 69.
15 As of the early 1960s the MARAD fleet contained the following:
    412 freighters ('reasonably up to date', including 312 Victory class ships, and 100 coastal dry cargo vessels)
        111 ships especially configured for military use
        70 tankers
        57 passenger ships
        74 ocean tugs
        833 'slow (10-knot), but still marginally serviceable' Liberty ships
        91 others
                                                    (Lawrence 1966: 114fn)
16 This point was emphasized to the author at a briefing on NATO sealift provided by officers of the US Military Sealift Command, Washington DC, 9 June 1982.
17 US Dept of Defense 1974, *Annual Report for FY 1975 and FY 1975-1980 Defense Program*, Washington DC: GPO: 165.
18 US Dept of Defense, *Annual Report for FY 1975*, op. cit.: 165.
19 Interviews with officials of the Senior Civilian Emergency Planning Committee, Brussels, 15 March 1983; White 1983: 32.
20 This account of the workings of the CSG is based upon the Military Sealift Command Briefing mentioned above.
21 Military Sealift Command Briefing, op. cit.
22 Interview with Admiral Isaac Kidd, USN (Ret.), Falls Church, Virginia, 23 March 1982; see also Kidd (1980).
23 An even more pressing problem was obtaining merchant ships to participate in NATO convoy exercises. During the Ocean Safari Exercise of 1977, four merchant ships then crossing the Atlantic were diverted from their normal routes for 24 hours in order to test tactical ASW. The cost of the diversion was nearly $200,000 (Wettern 1977: 25).
24 Sabotage by domestic political groups in the European NATO nations was not ruled out. Trade union based groups opposed to the mobilization have also been suspect, in that they might put container equipment out of action.
25 Witthoft (1975). References to this article are taken from a translation done for the author by Ms Kim McKeon.
26 In ports where the level of ocean traffic would be considered too small to justify an NCSO, civilian agencies might establish Reporting Offices. In the case of the United States, diplomatic officers might serve as Reporting Officers for shipping during an emergency. In addition, where there is already

a major allied NCSO, the US may simply have a Reporting Officer rather than establish an NCSO.
27 CP (Series I, Box 14, Folder 283), Letter from John R. Wadleigh to Richard G. Colbert, 19 August 1973.
28 For an account of this exercise, see US Dept of Defense *Current News*, Special Edition, Reforger 80, 26 November 1980.
29 'Exercise Strong Express in Retrospect', *International Defense Review*, December 1972: 661.
30 Interview with Admiral Harry Train, II, USN (Ret.), former SACLANT, 14 October 1983, Norfolk, VA.
31 UK, Ministry of Defence (1982: 26); see also the American assessment: US Dept of Defense (1983). The American report stressed the fact that the British ships did not have a great capability to unload where ports did not exist.

6 The USN, NATO and the war at sea

1 US Senate 1974: 88. Testimony by Chief of Naval Operations Admiral Elmo Zumwalt.
2 Ibid.: 95.
3 US Senate 1977: 74. Testimony of Secretary of Defense Harold Brown.
4 US Senate 1978: 1135. Testimony by Charles W. Duncan, Jr., Deputy Secretary of Defense.
5 Ibid.: 1255.
6 Interview, Admiral Harry Train, II, USN (Ret.), 14 October 1983, Norfolk, VA.
7 US Senate (1977); letter from Admiral Horacio Rivero, USN (Ret.), to author, 16 May 1983; interview with Admiral Gerald E. Miller, USN (Ret.), 22 April 1983, Arlington, VA.
8 NATO, Atlantic Command, Public Information Division, 'Statement of Admiral Harry D. Train II, USN, Commander-in-Chief US Atlantic Fleet to the Senate Armed Services Committee Preparedness Subcommittee, 17 February 1981 on Atlantic Fleet': 8.
9 The possibility that the movement of allied ASW forces into the Norwegian Sea might touch off a nuclear exchange is discussed in Posen (1982).

Bibliography

Acheson, Dean (1969) *Present At The Creation*, New York: W. W. Norton.
Alford, Jonathan (ed.) (1980) *Sea Power and Influence*, London: Gower.
Amme, C. H. (1967) *NATO Without France*, Stanford, California: Hoover Institute Press.
Arkin, William M. (1983) 'Nuclear Weapons At Sea', *Bulletin of Atomic Scientists*, Vol. 39, No 8, October.
Assembly of the Western European Union (AWEU) (1959) *State of European Security*, Report, Document No. 128, Paris, 20 May.
—— (1963) *State of European Security: Navies in the Nuclear Age*, Report, Document No. 269, Paris, 26 April.
—— (1967) *Defence of the Mediterranean and the NATO Southern Flank*, Report, Document No. 431, Paris, 4 December.
—— (1972) *Defence of the Northern and Southern Flanks*, Report, Document No. 568, Paris, 25 April.
—— (1973) *Security in the Mediterranean*, Report, Document No. 624, Paris, 7 November.
—— (1976) *Security in the Mediterranean*, Report, Document No. 708, Paris, 19 May.
Barnett, Roger W. (1987) 'The Maritime-Coalition Debate Isn't Over', *United States Naval Institute Proceedings (USNIP)*, Vol. 113, No. 6, June.
BDM Corporation (1977) *The Soviet Navy's Declaratory Doctrine for Theater Nuclear Weapons*, study prepared for the Director, Defense Nuclear Agency, Contract DNA-001-76-C-0230, Washington DC, 30 September.
Bender, Edward J. (1982) 'Naval Control of Shipping: A Reserve Commitment', *Naval Reserve Association Journal*, May.
Berman, Robert (1975) 'Soviet Naval Strength and Deployment' in M. MccGwire, K. Booth and J. McDonnell (eds) *Soviet Naval Policy: Objectives and Constraints*, New York: Praeger.
Bertram, Christoph and Holst, J. J. (1977) *New Strategic Factors in the North Atlantic*, Guildford, UK: IPC Science and Technology Press.
Blechman, Barry M. (1973) *The Changing Soviet Navy*, Washington DC: The Brookings Institution.
Blechman, Barry M. and Kaplan, Stephen S. (1978) *Force Without War: U.S. Armed Forces as a Political Instrument*, Washington DC: Brookings.

Bracken, Paul (1983) *The Command and Control of Nuclear Forces*, New Haven, Connecticut: Yale University Press.
Brassey's Defence Yearbook (1974), London: Brassey.
Breyer, Siefried (1970) *Guide to the Soviet Navy*, Annapolis, Maryland: Naval Institute Press.
Brodie, Bernard (1965) *A Guide to Naval Strategy*, New York: Praeger.
Brooks, Linton F. (1986) 'Naval Power and National Security: The Case for the Maritime Strategy', *International Security*, Vol. 11, No. 2, Fall.
Brown, Anthony C. (ed.) (1978) *Operation World War III, The Secret American Plan for War with the Soviet Union in 1957*, London: Arms and Armour Press.
Brown Harold (1983) *Thinking About National Security: Defense and Foreign in a Dangerous World*, Boulder, Colorado: Westview Press.
Cable, James (1981) *Gunboat Diplomacy*, New York, St Martin's Press.
—— (1983) *Britain's Naval Future*, Annapolis, MD: Naval Institute Press.
Canada, Department of Defense (1964) Directorate of History, *The RCN Today*, December 73/712.
—— (1968) Directorate of History, J. D. F. Kealy, *The Development of the Canadian Navy*, July, SSG II 223.
—— (1977) Directorate of History, J. D. F. Kealy, *Report of Exercise Ocean Safari 77*, 16–27 October, 78/44.
Carnegie Endowment for International Peace (1981) *Challenges for U.S. National Security*. 'Assessing the Balance: Spending and Conventional Forces', Washington DC: Carnegie Endowment for International Peace.
Case, Frank B. (1972) 'The Versatile, Vulnerable Container Ship', *USNIP*, Vol. 98, No. 2, February.
—— (1973) 'Time to Secure the Seas', *USNIP*, Vol.99, No.8, August.
Center for Defense Information (1981) *Nuclear Weapons in Europe: Documents*, Washington DC: Center for Defense Information.
Chichester, Michael and Wilkinson, John (1982) *The Uncertain Ally: British Defence Policy 1960–1980*, London: Gower.
Clark, John J. (1967) 'Merchant Marine and the Navy: A Note on the Mahan Hypothesis', *Royal United Services Institute Journal (RUSI)*, Vol. 112, No. 646, May.
Clawson, Carl H. (1980) 'The Wartime Role of Soviet SSBNs–Round Two', *USNIP*, Vol. 106, No. 3, March.
Cochran, Thomas B., Arkin, William M. and Hoenig, Milton M. (eds) (1984) *Nuclear Weapons Databook*, Vol. I, *U.S. Nuclear Forces and Capabilities*, Cambridge, Massachusetts: Ballinger Press.
Colletta, Paolo E. (1980a) *The American Naval Heritage in Brief*, 2nd Edition, Washington DC: University Press of America.
—— (1980b) *American Secretaries of the Navy*, Vol. II, *1913–1972*. Annapolis, MD: Naval Institute Press.
Cottrel, A. J. and Moorer, T. H. (1979) 'Seapower and NATO', a paper prepared for the Center for Strategic and International Studies conference on *NATO: The Second Thirty Years*, Brussels, Belgium, 1–3 September.
Crollen, Luc (1973) *Portugal, the U.S. and NATO*, Louvain, Belgium: Leuven University Press.
Daniel, Donald C. (1978) 'Trends and Patterns in Major Soviet Naval Exercises' in Paul J. Murphy (ed.) *Naval Power in Soviet Policy. Studies in Communist*

Affairs, Vol. 2 *The United States Air Force*, Washington DC: GPO (Government Printing Office).
Daniel, Marshall (1979) *Defense Transportation Organization: Strategic Mobility in Changing Times*, National Defense University Research Directorate, Memo-79-3, Washington DC: GPO.
Davis, Jacquelyn K. and Pfaltzgraff, Robert L. (1978) *Soviet Theater Strategy: Some Implications for NATO*, USSI Report 78-1, Washington DC: United States Strategic Institute.
Davis, Vincent (1966) *Post War Defense Policy and the U.S. Navy 1943-1946*, Chapel Hill, NC: University of North Carolina Press.
—— (1967) *The Admiral's Lobby*, Chapel Hill, North Carolina: University of North Carolina Press.
Davy, H. G. (1982) 'Merchant Shipping and the Maritime Threat', *Brassey's Defence Year book 1982*, London: Brassey.
de Cormoy, Guy (1970) *The Foreign Policies of France 1944-1968*, translated by Elaine P. Halperin, Chicago: University of Chicago Press.
Dennis, Michael F. H. (1971) *The Role of Navies in Limited War*, unpublished doctoral thesis, University of Minnesota.
Denny, Michael (1956) 'The Atlantic in a World War: What Does It Mean?', *RUSI*, Vol. 101, No. 603, August.
Dillon, G. M. (1972) *Canadian Naval Policy Since World War II: A Decision Making Analysis*, Occasional Paper, Dalhousie University, Halifax, Nova Scotia: Center for Foreign Policy Studies.
Dismukes, Bradford and McConnell, James (eds) (1979) *Soviet Naval Diplomacy*, Pergamon Policy Studies 37, New York: Pergamon Press.
Duncan, Charles K. (1972) 'The Maritime Equation: SACLANT in the 1970s', *NATO's Fifteen Nations*, Vol. 17, No. 1, February-March.
Dunn, Keith A. and Staudenmaier, William O. (1984) *Strategic Implications of the Continental Maritime Debate*, Washington Papers, Vol. XII, No. 107, New York: Praeger.
Dur, Philip A. (1976) *The Sixth Fleet: A Case Study of Institutional Naval Presence, 1946-1968*, unpublished doctoral thesis, Harvard University.
Enthoven, Alain G. and Smith, Wayne K. (1971) *How Much is Enough? Shaping the Defense Program 1961-1969*, New York: Harper and Row.
Erickson, John (1971) *Soviet Military Power*, London: Royal United Services Institute.
—— (1976) 'The Northern Theatre: Soviet Capabilities and Concepts', *Strategic Review*, Vol.3, No.2, Summer.
Finney, John W. (1975) 'US Assigns More Missile Submarines to the Defence of NATO', *New York Times*, 18 June.
Freedman, Lawrence (1982) 'The War of the Falkland Islands', *Foreign Affairs*, Vol. 61, No. 1, Fall.
Friedman, Norman (1980) 'SOSUS and US ASW Tactics', *USNIP*, Vol. 106, No. 3, March.
—— (1988) *The U.S. Maritime Strategy*, London: Janes.
Garde, Hans (1976) 'The Influence of Navies on the European Central Front', *USNIP*, Vol. 102, No. 5, May.
George, James L. (ed.) (1978) *Problems of Seapower as We Approach the Twenty-first Century*, Washington DC: American Enterprise Institute.

Gorshkov, S. G. (1979) *The Seapower of the State*, Annapolis, MD: Naval Institute Press.
Granger, Louis (1977) 'MSC Ships Sail in NATO Convoy', *Sealift*, Vol. 27, No. 11, November.
Gray, Colin S. (1986) *Maritime Strategy, Geopolitics and the Defense of the West*, New York: National Strategy Information Center.
Gretton, Peter (1965) *Maritime Strategy: A Study of British Defence Problems*, London: Cassell.
Grove, Eric J. (1987) *Vanguard to Trident: British Naval Policy Since World War II*, Annapolis, MD: Naval Institute Press.
—— (1989) 'The Norwegian Sea – NATO's First Line of Defence', in Eric Grove (ed.) *NATO's Defence of the North*, London: Brassey.
Guillon, Edward A. (ed.) (1968) *Uses of the Sea*, Englewood Cliffs, NJ: Prentice-Hall.
Hackett, John et al. (1978) *The Third World War: August 1985*, New York: Macmillan.
Hammond, Paul Y. (1963) 'Super Carriers and B-36 Bombers: Appropriations, Strategy and Politics' in Harold Stein (ed.) *American Civil Military Decisions, A Book of Case Studies*, Birmingham, Alabama: University of Alabama Press.
Hampshire, A. Cecil (1975) *The Royal Navy Since 1945*, London: William Kimber.
Hartley, Keith (1983) *NATO Arms Co-operation: A Study in Economics and Politics*, London: George Allen and Unwin.
Heymont, Irving (1966) 'The NATO Nuclear Bilateral Forces', *Orbis*, Vol. IX, No. 4, Winter.
Hezlet, Arthur (1967) *The Submarine and Sea Power*, London: Peter Davies.
Holloway, James L. (1983) 'Sealift', *USNIP*, Vol. 109, No. 6, June.
Hosti, Ole R., Hopmann, P., Terrance, P. and Sullivan, John D. (1973) *Unity and Disintegration in International Alliances: Comparative Studies*, New York: John Wiley & Sons.
Howe, Jonathan T. (1971) *Multicrises: Sea Power and Global Politics in the Missile Age*, Cambridge, Mass.: MIT Press.
Hunt, Ralph B. and Rosholdt, Erling F. (n.d.) *The Concepts of Notional Ship and Notional Ship Value in Logistic Capability Involving Merchant Ships*, The George Washington University Logistics Research Project, Washington DC, Contract Nr 761 (05), Project NR 047 001.
Huntington, Samuel P. (1954) 'National Policy and the Transoceanic Navy', *USNIP*, Vol. 80, No. 5, May.
Hurley, A. F. and Ehrhart, R. C. (eds) (1978) *Air Power and Air Warfare*, Proceedings of the 8th Military History Symposium 18–20 October 1978, Office of Air Force History, Headquarters USAF and US Air Force Academy, Washington DC: GPO.
International Institute for Strategic Studies (various years) *The Military Balance*, London: IISS.
—— (1976) *Power at Sea II: Superpowers and Navies*, Adelphi Papers, No. 123, London: IISS.
Jane's Fighting Ships (various years) London: Jane's Yearbooks.
Jantscher, Gerald R. (1975) *Bread Upon the Waters: Federal Aids to the Maritime Industries*, Washington DC: Brookings.

Jeschonnek, Gert (1956) 'The German Navy Within the NATO Defense System of Western Europe', paper prepared for the Command College, US Naval War College, Newport, RI.
—— (1969) 'Employment of Modern Naval Forces in the Defense of Northern and Central Europe', *U.S. Naval War College Review (NWCR)*, Vol. XXII, No. 3, November.
Karber, Philip A. (1970) 'Nuclear Weapons and Flexible Response', *Orbis*, Vol. XIV, No. 2, Summer.
Kaufman, William (1981) 'The Defense Budget' in Joseph A. Pechman (ed.) *Setting National Priorities: The 1982 Budget*, Washington DC: Brookings.
Kealy, J. D. F. and Russell, E. C. (1967) *A History of Canadian Naval Aviation, 1918–1962*, Ottawa: Queen's Printer.
Kennet, Lord (1964a) 'Navies in the Nuclear Age', *NATO Letter*, Vol. 12, Nos 2–4, February–April.
Kennet, Lord (1964b) 'Navies in the Nuclear Age', Part 2, *Western Europe, NATO Letter*, Vol. 12, No. 3, March.
Kidd, Isaac (1972) 'View from the Bridge of the Sixth Fleet Flag Ship', *USNIP* Vol. 92, No. 2, February.
—— (1978) 'Defense of the Atlantic', *NATO's Fifteen Nations*, Vol. 23, No. 5, October–November.
—— (1980) 'For Want of a Nail: The Logistics of the Alliance' in Kenneth Myers (ed.) *NATO: The Next Thirty Years*, Boulder, Colorado: Westview Press.
Kilmarx, Robert A. (1975) *United States–Soviet Naval Balance*, Report, Georgetown University Center for Strategic and International Studies (CSIS), Washington DC: CSIS, 1 April.
Kipp, Jacob W. (1977) 'Soviet Naval Aviation' in M. MccGwire and J. McDonnell *Soviet Naval Influence: Domestic and Foreign Dimensions*, New York: Praeger.
Komer, Robert (1984) *Maritime Strategy or Coalition Defense*, Lanham, Maryland: Abt Books.
Lawrence, Samuel A. (1966) *United States Merchant Shipping Policies and Politics*, Washington DC: Brookings.
Longworth, Brian (1983) 'The Case for a Maritime Strategy', *Defence*, Vol. XIV, No. 2, February.
Love, Robert W. Jr. (ed.) (1980) *The Chiefs of Naval Operations*, Annapolis, MD: Naval Institute Press.
Luttwak, Edward N. (1974) *The Political Uses of Sea Power*, Studies in International Affairs, No. 23, Baltimore, MD: The Johns Hopkins University Press.
MccGwire, M. K. (ed.) (1973a) *Soviet Naval Developments: Capability and Context*, New York: Praeger.
—— (1973b) 'Comparative Warship Building Programs' in M. MccGwire (ed.) *Soviet Naval Developments: Capability and Context*, New York: Praeger.
—— (1975) 'The Evolution of Soviet Naval Policy 1960–1974' in M. MccGwire, K. Booth and J. McDonnell (eds) *Soviet Naval Policy: Objectives and Constraints*, New York: Praeger.
—— (1980) 'The Rationale for the Development of Soviet Seapower', *USNIP*, Vol. 106, No. 5, May.
MccGwire, M.K. and McDonnell, John (eds) (1977) *Soviet Naval Influence, Domestic and Foreign Dimensions*, New York: Praeger.

MccGwire, M. K., Booth, Ken and McDonnell, John (eds) (1975) *Soviet Naval Policy: Objectives and Constraints*, New York: Praeger.
—— (1983) 'When Deterrence Fails: The Nasty Little War for the Falkland Islands', *NWCR*, Vol. XXXVI, No. 2, March–April.
McGruther, Kenneth R. (1978) *The Evolving Soviet Navy*, Newport, RI: Naval War College Press.
Mahan, Alfred T. (1957) *The Influence of Seapower Upon History 1660–1783*, New York: Hill and Wang.
Mako, William P. (1983) *U.S. Ground Forces and the Defense of Central Europe*, Washington DC: Brookings.
Marolda, Edward J. (1981) 'The Influence of Burke's Boys on Limited War', *USNIP*, Vol. 107, No. 8, August.
Marriott, John (1968) 'NATO and the Mediterranean', *Navy*, Vol. 73, No. 12, December.
—— (1973) 'Exercise Strong Express', *NATO's Fifteen Nations*, Vol. 18, No. 2, February–March.
Martin, Lawrence W. (1965) *The Sea in Modern Strategy*, London: Chatto & Windus.
Martin, Lawrence and Bull, Hedley (1968) 'The Strategic Consequences of Britain's Revised Naval Roles' in Edward A. Guillon (ed.) *Uses of the Seas*, Englewood Cliffs, NJ: Prentice Hall.
Mayer, R. Charles and Handerson, Harold (1974) 'A Critique of the Rationales For Present U.S. Maritime Programs', *Transportation Journal*, Winter.
Mearsheimer, John J. (1986) 'A Strategic Misstep: The Maritime Strategy and Deterrence in Europe', *International Security*, Vol. 11, No. 2, Fall.
Mills, Walter (1951) 'Seapower: Abstraction or Asset?', *Foreign Affairs*, Vol. 29, No. 3, April.
Morse, John H. (1977) 'Questionable NATO Assumptions', *Strategic Review*, Vol. 5, No. 1, Winter.
Moulton, J. L. (1974) 'NATO and the Atlantic', *Brassey's Defence Yearbook*, London; Brassey.
Mountbatten, Louis (1955) 'Allied Naval and Air Commands in the Mediterranean', *RUSI*, Vol. C, No. 598, May.
Mulley, F. W. (1962) *The Politics of Western Defense*, New York: Praeger.
Murphy, Paul J. (ed.) (1978) *Naval Power in Soviet Policy*, Studies in Communist Affairs, Vol. 2, United States Air Force, Washington DC: GPO.
Myers, Kenneth (1979) *North Atlantic Security: The Forgotten Flank?* The Washington Papers, Vol. VI, No. 62, Beverly Hills and London: Sage Publications.
—— (ed.) (1980) *NATO: The Next Thirty Years*, Boulder, Colorado: Westview Press.
National Defense University (1978) *Equivalence, Sufficiency and the International Balance*, Proceedings of the 5th National Security Affairs Conference, Washington DC: National Defense University.
Nitze, Paul H., Sullivan, Leonard and the Atlantic Council Working Group on Securing the Seas (1979) *Securing the Seas: The Soviet Naval Challenge and Western Alliance Options*, Boulder, Colorado: Westview Press.
North Atlantic Assembly (1982) *NATO Anti-Submarine Warfare: Strategy, Re-*

quirements and the Need for Co-operation; Report of the Sub-Committee on Defense Co-operation, Brussels.
North Atlantic Treaty Organization (NATO) (1955) *NATO: The First Five Years, 1949–1954*, Report by Lord Ismay, Paris: NATO.
—— (1969) Supreme Allied Commander Atlantic, *Report of Study: Relative Maritime Strategies and Capabilities of NATO and the Soviet Bloc*, The Brosio Study, Norfolk, Virginia, March.
—— (1981a) Address by the Supreme Allied Commander, Atlantic, to the Atlantic Council of Canada, November, Headquarters Supreme Allied Command, Atlantic.
—— (1981b) *NATO: Facts and Figures*, Brussels: NATO Information Services.
—— (1982) *Texts of Final Communiques* (Vol. I, 1949–1974, Vol. II, 1975–1980), Brussels: NATO Information Services.
Ørvik, Nils (1963) *Europe's Northern Cap and the Soviet Union*, Occasional Papers on International Relations, No. 6, Center for International Affairs, Harvard University, Cambridge, Mass.: CFLA.
Osgood, Robert E. (1962) *Nuclear Control in NATO*, Washington DC: Washington Center for Foreign Policy Research.
—— (1964) *The Case for the MLF: A Critical Evaluation*, Washington DC: Washington Center for Foreign Policy Research.
—— (1968) *Alliances and American Foreign Policy*, Baltimore MD: The Johns Hopkins University Press.
Palmer, Michael A. (1988) *The Origins of the Maritime Strategy*, Washington DC: Department of the Navy, Naval Historical Center.
Pechman, Joseph A. (ed.) (1981) *Setting National Priorities: The 1982 Budget*, Washington DC: Brookings.
Polmar, Norman (1972) *Soviet Naval Power*, New York: National Strategy Information Center.
Poole, Walter (n.d.) 'Organizing NATO's Naval Commands: Anglo-American Rivalry', unpublished study.
Posen, Barry R. (1982) 'Inadvertent Nuclear War? Escalation and NATO's Northern Flank?', *International Security*, Vol. 7, No. 2, Fall.
Pranger, Robert J. and Labrie, Roger P. (eds) (1977) *Nuclear Strategy and National Security: Points of View*, Washington DC: American Enterprise Institute.
Ranft, Bryan and Till, Geoffrey (1983) *The Sea in Soviet Strategy*, Annapolis, MD: Naval Institute Press.
Rapping, L. A. (1963) *Federal Maritime Policy and Military Shipping Requirements*, Rand Memorandum RM-3422-ISA, Prepared for the office of Assistant Secretary of Defense for International Security Affairs, Santa Monica, California: The Rand Corporation, April.
Record, Jeffrey (1974) *U.S. Nuclear Weapons in Europe: Issues and Alternatives*, Washington DC: Brookings.
Richardson, Robert C. (1981) 'NATO's Nuclear Strategy: A Look Back', *Strategic Review*, Vol. 9, No. 2, Spring.
Richmond, Herbert W. (1928) *National Policy and Naval Strength* and other essays, London: Green & Co.
Rivero, Horacio (1972) 'The Defence of NATO's Southern Flank', *RUSI*, Vol. 117, No. 666, June.

—— (1978) *Reminiscences of Horacio Rivero Jr*, Annapolis, MD: Naval Institute Press.
Robbins, James J. (1955) *Recent Military Thought in Sweden on Western Defense*, Rand Research Memorandum 1407, Santa Monica, California: The Rand Corporation, 25 January.
Roberts, Stephen S. (1981) 'Western European and NATO Navies', *USNIP*, Vol. 107, No. 3, March.
Rohwer, Jurgen (1975) *Superpower Confrontation on the Seas*, The Washington Papers, Vol. III, No. 26, Beverly Hills and London: Sage Publications.
Rosenberg, David Allan (1976) 'The U.S. Navy and the Problem of Oil in a Future War: The Outline of a Strategic Dilemma, 1945–1950', *NWCR*, Vol. XXIX, No. 1, Summer.
—— (1978) 'American Postwar Air Doctrine and Organization: The Navy Experience' in A. F. Hurley and R. C. Ehrhart (eds) *Air Power and Air Warfare*, Washington DC: GPO
—— (1981/2) 'A Smoking Radiating Ruin at the End of Two Hours: Documents on American Plans for Nuclear War with the Soviet Union, 1954–1955', *International Security*, Vol. 6, No. 3, Winter.
Roskill, Stephen (1962) *The Strategy of Sea Power. Its Development and Application*, London: Collins.
Sanginetti, T. (1965) *Atome et Bataille sur la Mer*, Paris: Hachette.
Schlesinger, J. R. (1977) 'The Theater Nuclear Force Posture in Europe' in R. J. Pranger and R. Labrie (eds) *Nuclear Strategy and National Security: Points of View*, Washington DC: American Enterprise Institute.
Schratz, P. R. (1980) 'Paul Henry Nitze' in Paolo E. Colletta (ed.) *American Secretaries of the Navy*, Vol. II, Annapolis, Maryland: Naval Institute Press.
Simons, William E. (1984) 'U.S. Reinforcement of NATO', *Journal of Defense and Diplomacy*, Vol. 2, No. 8, August.
Sokol, Anthony E. (1961) *Seapower in the Nuclear Age*, Washington DC: Public Affairs Press.
Sokolsky, Joel J. (1984) 'The United States Navy and Nuclear ASW Weapons', *USNIP*, Vol. 110, No. 12, December.
—— (1986) *Seapower in the Nuclear Age: NATO As A Maritime Alliance*, unpublished doctoral thesis, Harvard University.
—— (1989) 'Anglo-American Maritime Strategy in the Era of Flexible Response, 1960–1980', in John B. Hattendorf and Robert S. Jordan (eds), *Maritime Strategy and the Balance of Power: Britain and America in the Twentieth Century*, London: St. Anthony's/Macmillan.
Sorrels, Charles A. (1983) *U.S. Cruise Missile Programs: Development, Deployment and Implications for Arms Control*, New York: McGraw-Hill.
Stein, Harold (ed.) (1963) *American Civil-Military Decisions: A Book of Case Studies*, Birmingham, Alabama: Alabama University Press.
Stockholm International Peace Research Institute (1970) *SIPRI Yearbook of World Armaments and Disarmament*, 1969/70, Stockholm: Almqvist and Wiksell.
—— (1978) *Tactical Nuclear Weapons: European Perspectives*, New York: Crane Russak.
—— (1979) *SIPRI Yearbook of World Armaments and Disarmament*, 1979, London: Taylor and Francis.

The Sunday Times of London Insight Team (1982) *War in the Falklands*, London: Times Newspapers.
Swartztrauber, Sayre A. (1979) 'The Potential Battle of the Atlantic', *USNIP*, Vol. 105, No. 5, May.
Till, Geoffrey et al. (1982) *Maritime Strategy and the Nuclear Age*, New York: St Martin's Press.
Train, Harry (1982) 'Challenge At Sea: Naval Strategy for the 1980s', *NATO's Fifteen Nations*, Vol. 27, Special Edition 2.
Troost, Admiral Carlisle A. H. (1987) 'Looking Beyond the Maritime Strategy', *USNIP*, Vol. 113, No. 1, January.
Underwood, John L. (1979) *Conflict in the Eastern Mediterranean*, Center for Naval Analyses, Memorandum, Alexandria, Virginia: Center for Naval Analyses, 20 August.
United Kingdom, Ministry of Defence (1978) *NATO – The British Contribution to allied Defence*, London, HMSO, April.
―― (1981) Secretary of State for Defence, *Statement on the Defence Estimates 1981*, London: HMSO.
―― (1982) Secretary of State for Defence, *The Falklands Campaign: The Lessons*, London, HMSO.
United States Comptroller General of the United States (1976) *The National Defense Reserve Fleet: Can It Respond to Future Contingencies?* Washington DC: GPO.
United States Congress (1959) House of Representatives. Committee on Armed Services, Hearings and Report. *Adequacy of Transportation Systems in Support of the National Defense Effort in the Event of Mobilization*, 86th Cong., 1st Sess., Washington DC: GPO.
―― (1962) House of Rep. Committee on Merchant Marine and Fisheries. Hearings. *Review of Merchant Marine Policy 1962*. 87th Cong., 2nd Sess., Washington DC: GPO.
―― (1965) House of Rep. Committee on Merchant Marine and Fisheries. Hearings. *Operation Steel Pike I*, 89th Cong., 1st Sess., Washington DC: GPO.
―― (1966) House of Rep. Committee on Merchant Marine and Fisheries. Hearings. *Vietnam Shipping Policy Review*, Part II. 89th Cong., 2nd Sess., Washington DC: GPO.
―― (1973) House of Rep. Subcommittee on Military Applications, Joint Committee on Atomic Energy. Hearings. *Military Applications of Nuclear Technology*, Part 2, 93rd Cong., 1st Sess., Washington DC: GPO.
―― (1980a) House of Rep. Committee on Merchant Marine and Fisheries. Hearings. *Defense Sealift Capability*. 95th Cong., 2nd Sess., Washington DC: GPO.
―― (1980b) House of Rep. Committee on Merchant Marine and Fisheries. Hearings. *Defense Sealift Capability and Merchant Ship Attrition*, 96th Cong., 2nd Sess., Washington DC: GPO.
United States Congressional Budget Office (CBO) (1967a) *Planning U.S. General Purpose Forces: The Navy*, Budget Issue Paper, Washington DC: CBO.
―― (1976b) *U.S. Naval Forces Alternatives*, Staff Paper by Dov Zakheim, Washington DC: CBO.
―― (1977) *The U.S. Sea Control Mission: Forces, Capabilities and Requirements*, Washington DC: CBO.

—— (1979) *U.S. Airlift Forces: Enhancement Alternatives for NATO and Non-NATO Contingencies*, Washington DC: CBO.
—— (1980) *Shaping the General Purpose Navy of the Eighties: Issues for Fiscal Years 1981–1985*, Washington DC: CBO.
United States Department of Commerce, Maritime Administration, *Annual Reports*, Washington DC: GPO.
United States Department of Defense. Office of the Sec. of Def. *Annual Reports*, Washington DC: GPO.
—— (n.d.) Dept of the Air Force, Strategic Air Command. *History of the Joint Strategic Target Planning Staff: Background and Preparation on SIOP-62*, Omaha, Nebraska: SAC History and Research Division.
—— (1950) Dept of the Navy, Office of the Chief of Naval Operations. *Project Hartwell: A Report on the Security of Overseas Transport*, prepared by the Massachusetts Institute of Technology, Cambridge, Mass., September.
—— (1952) Dept of the Navy, *All Hands*, No. 428, October.
—— (1954) Office of the Sec. of Def. *Memorandum of Agreement Between the Department of Defense and the Department of Commerce Dealing with the Utilization, Transfer and Allocation of Merchant Ships*, The Wilson–Weeks Agreement, Washington DC: July.
—— (early 1960s) Dept of the Navy, Office of the Chief of Naval Operations. *War-at-Sea Studies*, prepared over several years.
—— (1965) President's Maritime Advisory Committee. *Maritime Policy and Programs of the United States*, Washington DC: GPO.
—— (1966/1967) Dept of the Navy, Office of the Chief of Naval Operations. *Sealift Requirements Study*, Part I, Washington DC, August 1966; Part II, Washington DC, February 1967.
—— (1972a) Office of the Sec. of Def., Assistant Sec. of Def., Program and Analysis. *Sealift Procurement and National Security*, 'SPANS' study, Washington DC: August.
—— (1972b) Office of the Assistant Sec. of Def. (Public Affairs) *News Release*, 1 August.
—— (1973) Organization of the Joint Chiefs of Staff. *Allied Guide to Masters*, Washington DC: December.
—— (1976) Office of the Sec. of Def. *Report to Congress on U.S. Conventional Reinforcement to NATO*, Washington DC: June.
—— (1977) Defense Nuclear Agency. The *Soviet Navy Declaratory Doctrine for Theatre Nuclear Warfare*. Completed by the BDM Corporation, Washington, DC.
—— (1979a) Organization of the Joint Chiefs of Staff. Condit, Kenneth W. *The History of the Joint Chiefs of Staff: The Joint Chiefs of Staff and National Policy*, Vol. II, 1947–49, Washington DC: Historical Division, JSC Secretariat.
—— (1979b) Organization of the Joint Chiefs of Staff. Poole, Walter S. *The History of the Joint Chiefs of Staff: The Joint Chiefs of Staff and National Policy*, Vol. IV, 1950–52, Washington DC: Historical Division, JSC Secretariat.
—— (1980) Office of the Sec. of Def. *An Evaluation Report of Mobilization and Deployment Capability Based on Exercises Nifty Nugget 78 and Rex 78*, Washington DC: 30 June.
—— (1981) Office of the Sec. of Def. *Memorandum of Agreement Between the*

Bibliography 213

Department of Defense and the Department of Commerce on Procedures for Shipping Support of Military Operations, Washington DC: June.

—— (1983) Dept of the Navy, Office of the Secretary of the Navy. *Lessons of the Falklands: Summary Report*, Washington DC: February.

—— (1984) Dept of the Navy, text of statement by Admiral Kinnard R. McKee, USN, Director Naval Propulsion Program, before the Subcommittees on Research and Development and Seapower and Strategic Minerals of the United States House of Representatives, Armed Services Committee, 6 February.

United States General Accounting Office (GAO) (1959) *Review of the National Defense Reserve Fleet for Fiscal Years 1957 and 1958*, Washington DC: GPO.

United States Library of Congress, The Papers of Admiral Lynde McCormick (MP).

United States National Archives, Washington DC. The Records of the Joint Chiefs of Staff (RJ).

United States National Research Council (NRC) (1959) Panel on the Wartime Use of the U.S. Merchant Marine. *The Role of the U.S. Merchant Marine in National Security*. Project Walrus, Washington DC: National Academy of Sciences.

United States Senate (1973) Committee on Foreign Relations, Report, *U.S. Security Issues in Europe: Burden-Sharing and Offset, MBFR and Nuclear Weapons*, 93rd Cong., 1st Sess, Washington DC: GPO.

—— (1974) Committee on Appropriations, Hearings. *Department of Defense Appropriations for Fiscal Year 1975*, Part 3, The Navy, 93rd Cong., 2nd Sess., Washington DC: GPO.

—— (1976) Committee on Commerce, Committee Print. *Soviet Oceans Development*, 94th Cong., 2nd Sess., Washington DC: GPO.

—— (1977a) Committee on Armed Services, Hearings. *NATO Posture and Initiatives*, 95th Cong., 1st Sess., Washington DC: GPO.

—— (1977b) Committee on Armed Services, Report. *U.S. Naval Forces in Europe*, 95th Cong., 2nd Sess., Washington DC: GPO.

—— (1978) Committee on Armed Services, Hearings. *Department of Defense Authorization for Appropriations for Fiscal Year 1979*, Part 2, 95th Cong., 2nd Sess., Washington DC: GPO.

United States, US Naval War College, Naval Historical Collection, The Richard G. Colbert Papers (CP).

United States, US Navy, Operational Archives, Washington DC. Chief of Naval Operations Files.

—— Operational Archives, Washington DC. War Plans.

—— Operational Archives, Washington DC. Strategic Plans.

—— Operational Archives, Washington DC. The Papers of Admiral Sherman (SP).

—— Operational Archives, Washington DC. Strategic Plans, Miscellaneous Plans and Studies.

—— Operational Archives, Washington DC. Short of War Documentation, Berlin File.

—— (1946a) Operational Archives, Washington DC. Command File, Post 1 January. Annual Reports of Commander-in-Chief Atlantic Fleet.

—— (1946b) Operational Archives, Washington DC. Command File, Post 1 January. Annual Reports of Commander-in-Chief US Naval Forces Europe.

—— (1946c) Operational Archives, Washington DC. Command File (CF), Post

1 January. Annual Reports of Commander-in-Chief US Naval Forces, Eastern Atlantic and Mediterranean.
—— (1947) Operational Archives, Washington DC. War Plans.
—— (1948) Operational Archives, Washington DC. The General Board, *National Security and Naval Contributions Thereto*, 425, Serial 315, June.
—— (1950) Operational Archives, Washington DC. Special Study by Vice Admiral Low, *The Low Report*, April.
—— (1963) Operational Archives, Washington DC. Short of War Documentation, Special List No. 3, Keeping the Peace Chronology, August.
—— (1966) Operational Archives, Washington DC. Short of War Documentation, *The Navy and cr Conflicts*, study by the Bendix Corporation, September.
—— (1977) Operational Archives, Washington DC. Short of War Documentation, *U.S. Navy Responses to International Incidents and Crises, 1955–1975*. Vol. II, by Robert B. Mahoney, Center for Naval Analysis.
Vlahos, Michael (1982) 'Maritime Strategy vs. Coalition Commitment', *Orbis*, Vol. 26, No. 3, Fall.
Wallin, Lars B. (ed.) (1982) *The Northern Flank in a Central European War*, Stockholm: The Swedish National Defence Research Institute.
Watkins, Admiral James D. (1986) 'The Maritime Strategy', *USNIP* (Special Supplement) January.
Watson, Bruce (1982) *Red Navy at Sea: Soviet Naval Operations on the High Seas*, Boulder, Colorado: Westview Press.
Watson, Bruce W. and Walton, M. A. (1976) 'Okean 75', *USNIP*, Vol. 102, No. 7, July.
Wegner, Edward (1975) *The Soviet Naval Offensive*, Annapolis, MD: The Naval Institute Press.
Weinland, Robert G. (1975) 'The State and Future of the Soviet Navy in the North Atlantic' in M. MccGwire, K. Booth and J. McDonnell (eds) *Soviet Naval Policy: Objectives and Constraints*, New York: Praeger.
—— (1980) *Northern Waters: Their Strategic Significance*. Professional Paper, 328, Center for Naval Analyses, Alexandria, VA: Center for Naval Analysis, December.
West, F. J. (1979) 'U.S. Naval Forces and the Northern Flank of NATO', *NWCR*, Vol. XXXII, No. 4, July–August.
Wettern, Desmond (1977) 'Ocean Safari 1977', *Navy International*, Vol. 82, No. 12, December.
—— (1981) 'Defended Lanes vs. Convoys', *Navy International*, Vol. 86, No. 12, December.
White, Cliff (1983) 'Reinforcing Reassurance and Deterrence', *RUSI*, Vol. 128, No. 3, September.
Whitely, Peter (1979) 'The Reinforcement of Europe', *NATO's Fifteen Nations*, vol. 24, No. 4, August–September.
Wilmott, H. P. (1981) *Sea Warfare: Weapons, Tactics and Strategy*, Strettington, Chichester, UK: Anthony Bird.
Wilson, Desmond P. (1976) *The U.S. Sixth Fleet and the Conventional Defense of Europe*, Professional Paper No. 160, Arlington, VA: Center for Naval Analyses, September.
Wilson, Desmond P. and Brown, Nicholas (1971) *Warfare at Sea: Threat of the*

Seventies, Professional Paper, No. 79, Arlington, VA: Center for Naval Analyses, November.
Winton, John (1983) *Convoy: The Defence of Sea Trade 1890–1990*, London: Michael Joseph.
Witthoft, Hans J. (1975) 'Handelsschiffahrt im Verteidigungsfall', *Marine-Rundschau*, Vol. 72, No. 9, September.
Wolfe, Thomas (1975) *Military Power and Soviet Policy*, Rand Memorandum, No. P5388, Santa Monica, California: The Rand Corparation.
Zakheim, Dov Z. (1978a) 'Maritime Presence, Projection and the Constraints of Parity', *Equivalence, Sufficiency and the International Balance*, Washington DC: National Defense University.
—— (1978b) 'Land-based Aircraft Options for Sea Control' in James L. George (ed.) *Problems of Sea Power As We Approach the Twenty-First Century*, Washington DC: American Enterprise Institute.
—— (1982) 'The Power Balance in the North Atlantic' in Lara B. Wallin (ed.) *The North Flank in a Central European War*, Stockholm: The Swedish National Defence Research Institute.
Zimmerman, Peter D. and Greb, Allen (1982) 'The Bottom Rung of the Ladder: Battlefield Nuclear Weapons in Europe', *NWCR*, Vol. XXV, No. 6, November–December.
Zumwalt, Elmo R. (1976) *On Watch*, New York: Quadrangle/The New York Times Book Co.

BRIEFINGS

Supreme Allied Command Atlantic, Norfolk, Virginia, February 1982, February 1983, October 1983.
US Military Sealift Command, Washington DC, June 1982.
NATO Headquarters, Civil Emergency Planning Staff, Brussels, March 1983.
Supreme Headquarters Allied Powers Europe, Mons, March 1983.
Ministry of Defence, London, 23 March 1983.
Headquarters, Allied Command Channel, Northwood, March 1983.
National Defence Headquarters, Operational Research and Analysis Establishment, Ottawa, November 1983.

INTERVIEWS

Admiral Jerauld Wright, USN. ret., Washington DC, March 1981.
The Honorable Paul H. Nitze, Arlington, VA, July 1981.
Admiral Elmo Zumwalt, USN. ret., Arlington, VA, March 1982.
Admiral Isaac Kidd, Jr., USN. ret., Falls Church, VA, March 1982.
Mr Frank Case, Maritime Administration, Washington DC, October 1982.
Rear Admiral Sayre Swartztrauber, USN, Washington DC, January 1983.
Professor Michael MccGwire, Washington DC, January 1983.
Admiral J. H. Scheuer, Royal Dutch Navy, Brussels, March 1983.
Professor Geoffrey Till, Royal Naval College, Greenwich, March 1983.
Admiral of the Fleet, The Lord Hill-Norton, London, March 1983.
Admiral Paul de Cazanove, French Navy, ret., Paris, March 1983.
Admiral Gerald E. Miller, USN. ret., Arlington, VA, April 1983.
Admiral Harry Train, USN. ret., Norfolk, VA, October 1983.
Commodore Fred J. Mifflin, Maritime Command, Halifax, March 1984.

Index

Acheson, Dean 17
ACTIVITY 16
Aegean Sea 32, 35, 110
Africa 83, 101
Algeria 50, 57
Allied Command Europe (ACE) 64, 98
Allied Forces Mediterranean (AFMED) 30, 32, 35–6, 42, 45, 49–52, 57, 109
Allied Naval Communications Agency (ANCO) 57
anti-aircraft missiles (SAM) 76
anti-ship missiles (SSM) 76, 81–2, 116
anti-submarine warfare (ASW) 1, 4, 8, 11, 16–17, 21, 23, 28–9, 44–5, 48–50, 54–5, 57, 61, 65–6, 68, 73, 88, 99, 110, 114–19, 121, 168, 181, 183, 185
Argentina 166
Atomic Energy Commission 58
Azores 8, 13, 51, 57
Azores Fixed Acoustic Range (AFAR) 119

Baltic Sea 38–42, 87, 89, 91, 100, 162–4, 167
Baltysk 39
Barcelona 22
Barents Sea 13, 81, 85, 88–9
Bay of Biscay 51, 155–6
BEEHIVE I 17
Belgium 8, 13–14, 17, 36, 48, 98, 112, 118–19, 167, 172, 180; Navy 48
Berlin Crisis 33, 61, 66–7, 74, 138, 154
Bertram, Christoph 4
BIG LIFT 138–9
Black Sea 12, 32, 35, 39, 110
Bradley, General Omar 73
Brazil 166
Breyer, Siefried 90
Brosio, Manlio 112
Brosio Study 113–16, 177
Brown, Harold 179, 184
Brown, Nicholas 100
BUFFET 21
Burke, Admiral Arleigh 67

Canada 13–14, 28–9, 54–5, 98, 118–19, 130; Royal Canadian Air Force (RCAF) 18; Royal Canadian Navy (RCN) 11, 21, 54–5
Carnegie Endowment Study (1981) 183
Carney, Admiral Robert 17, 28–9
Casablanca 22
Center for Naval Analyses Study (1959) 61
Center for Naval Analyses Study (1971) 100
Cherbourg 22
Churchill, Winston 17, 63
Civil Air Reserve Fleet (CRAF) 155
Civilian Sealift Group (CSG) 152, 155, 157–9
Colbert, Richard G. 102–4, 117, 166

Index 217

collective defense 14, 45, 51, 103–6, 109, 111, 126, 130, 174, 186
Commander-in-Chief Allied Forces North (CINCNORTH) 39–41, 43, 48, 56
Commander-in-Chief Allied Forces South (CINCSOUTH) 29–30, 32, 35–6, 43, 52, 60–2, 96, 109, 111, 166
Commander-in-Chief Atlantic (CINCLANT) 9, 21–2, 27, 30
Commander-in-Chief Channel (CINCHAN) 4, 29, 36–7, 40, 43–4, 46, 51, 111, 118, 120
Concept of Maritime Operations (CONMAROPS) 2, 120–1, 189–90
containerization 160–1, 168, 173
conventional deterrence 3–6, 11, 46–8, 59, 70, 73, 92, 94–7, 100–2, 113, 115, 117, 122, 125, 130, 138, 146, 153, 158, 167, 170, 177, 180, 183–4, 186–9
CRESCENT MACE 60
Crete 120
Cuba 83, 101
Cuban Missile Crisis 55, 61, 83
Cyprus 33, 110

Daisy Chain 60
Danzig 39
Davis Strait 55
Davy, H. G. 173
DAWN PATROL 121
Defence Shipping Authority (DSA) 128–9, 131–4, 136, 152, 158–9, 161, 163, 174
Defense Planning Committee 64, 106, 109, 111, 116, 119, 145
De Gaulle, Charles 50–1
Denfeld, Admiral Louis 15–16
Denmark 8, 13–14, 16–17, 20–1, 39–40, 42, 92, 112, 119, 167; Air Force 40; Navy 40, 48
Denny, Sir Michael 126, 133
deterrence 14, 32–3, 44–5, 95–7, 99, 102, 113, 123, 170, 184, 187, 189
DROPSHOT 12
Duncan, Admiral Charles 109

Egypt 83
Eilat destroyer 82
Eisenhower, General Dwight 15, 24, 26, 29, 58–9
Emergency Defense Plan (EDP 1–52) 23, 25–7
EMIGRANT 21
Erickson, John 87
ESTABLISH CONTACT 97
Evaluation of the Atomic Offensive (1955) 73

Falklands War 173
flexible response 41–2, 69–71, 75, 92–7, 99, 100–2, 104, 109, 111–12, 114, 117, 121–2, 138, 145–6, 150, 159, 161, 169, 172, 179, 187–90
forward defense 120–1, 130, 187
France 8, 13–14, 16–17, 20, 29, 30, 32, 35–6, 48–53, 57–8, 63, 100, 110, 119, 122, 147–8, 156, 181; Navy 50–2, 56–7, 116, 119

GANNETT 21
General Strike Plan (GSP) 64–5
Germany 6, 8, 21, 39, 42, 63, 98, 112, 118–19, 155, 163–4, 167, 172, 174, 180; Navy 39–40, 48–9
Gibraltar 33, 35, 51, 110
Gorshkov, Admiral Sergei 86
Greece 29–30, 32, 35, 45, 96, 109, 111, 119, 121, 160; Navy 49, 56
Greenland 9, 13, 89
Greenland–Iceland–United Kingdom gap (GIUK) 56, 83, 91, 115, 180–1, 183
Gretton, Sir Peter 71
Grove, Eric 117, 120

Harmon Report 26
Hill-Norton, Lord 68, 90
Holloway, James 179

Iberia 11
Iceland 8–9, 12–13, 57, 89, 98–9
intercontinental ballistic missiles (ICBM) 3

218   Index

intermediate-range ballistic missiles (IRBM) 61
Iron Curtain 8
Ismay, Lord 127–8
Italy 8, 14, 17, 29–30, 32, 35, 51, 57, 61–2, 96, 98, 109, 111, 119, 121; Navy 29, 49, 56–7, 89

Jeschonnek, Admiral Gert 40–1
Johnson, Nicolas 138
Joint Chiefs of Staff (JCS) 9, 12–13, 22, 24, 26–30, 63, 117, 129–32, 136, 144, 147, 150–1, 157–9, 166; Joint Military Transportation Committee (JMTC) 132; Joint Strategic Plans Committee (JSPC) 25
Jordanian Crisis 82
Jutland peninsula 36, 40

Kauffman, William 171
Kennedy, John F. 63
Kennedy Administration 71, 73, 137
Kidd, Isaac 111, 167, 183
Kola Peninsula 83, 185
Komer, Robert 150
Konigsberg 39
Korea 8
Korean War 14, 126, 129, 134, 139–40
Kronstadt 39

Lawrence, Samuel A. 137
Le Havre 22
LIFE LINE 29
Lisbon 24, 72
Long Term Defense Program (LTDP) 119
Low Report 10
Luxembourg 98, 172

Madeira 13
Mahan, Alfred Thayer 1–2, 7, 67, 125, 190
MAINBRACE 21
Malta 32–3, 110–11
MARINER 28
Maritime Administration (MARAD) 133–5, 149–50, 152, 158, 162

Maritime Contingency Force (MCF) 104, 109, 111, 113
Marseilles 22
Martin, Lawrence 67, 71
MATCHMAKER 57, 104
MccGwire, Michael 76, 84–6, 90
McCormick, Admiral Lynde 18, 20–2, 28, 54
McGruther, Kenneth R. 91
McNamara, Robert S. 73, 117, 137–9
MEDFLEXABLE 35
Medium Term Defence Plan 13, 132
Merchant Shipping Control Zones (MERZONE) 162, 164
Middle East War (1973) 82
Military Airlift Command (MAC) 154
Military Standing Committee 13, 20, 57, 93, 105–6, 112, 130, 161
Mills, Walter 16
Montgomery, Field Marshall Bernard 58
Mountbatten, Lord Louis 17, 30, 32, 34–5
Multilateral Force (MLF) 62–4
Murmansk 12, 35

National Defense Reserve Fleet (NDRF) 132–7, 144–5, 148–9, 155; Ready Reserve Force (RRF) 148–9, 154
National Security Council (NSC) 135
National Shipping Authority (NSA) 134
NATO Naval Armaments Group (NNAG) 119
Netherlands 8, 13–14, 16–17, 20–1, 36, 48–9, 98, 104, 117, 119, 155, 167, 172, 180; Navy 48
NEW BROOM 28
NEW BROOM II 28
Newport Study 117
NIFTY NUGGET 122, 172
Nimitz, Admiral Chester 8
Nitze, Paul H. 67, 183
Norstad, General Lauris 72
North Africa 8
North Atlantic Council 32, 35, 44, 51, 58, 89, 93, 102, 104, 112, 127,

Index 219

130, 143, 150–1, 153, 158, 162, 175
North Sea 18, 24, 38–41, 46, 83, 87, 118, 121–2, 164, 167, 180
Norway 13–14, 16–17, 21, 46, 49, 85, 87, 89, 94, 97–8, 100, 104, 112, 115–16, 119–20, 164, 179–80, 184; Navy 48–9
Norwegian Sea 13, 41–2, 71, 81, 85–9, 115–16, 118, 121–2, 179–80, 183, 190
nuclear attack submarine (SSN) 76, 82, 87
nuclear cruise missile attack submarine (SSGN) 76, 82, 87, 115
Nuclear Planning Group (NPG) 64
nuclear-powered ballistic missile submarine (SSBN) 1, 3, 47, 50, 54, 61–3, 70, 76, 81, 83–5, 87–8, 91, 97, 100, 103, 114–15, 121–2, 183, 186–7, 190

OCEAN SAFARI 168
OCEAN SAFARI (77) 121
OFFTACKLE 11
OKEAN (70) 83–4
OKEAN (75) 83–4
Operation Control Authority (OCA) 163
OPERATION TEAMWORK 51
OPLAN (2300) 69
Oran 22
Ørvik, Nils 126
Osgood, Robert 63

Persian Gulf 87, 122, 149
Planning Board for Ocean Shipping (PBOS) 22, 127–9, 132–3, 135, 150–1, 158, 162–3, 170; see also SCEPC
Portugal 13, 21, 49, 57, 151, 167–8; Air Force 49; Navy 49
Priority Strike Program (PSP) 64–5
PROGRESS 16
Project Hartwell 65

Quick Reaction Alert (QRA) 65, 96

Rand Corporation Study (1962) 126–7
REFORGER 122, 167
Richardson, General Robert C. III 72
Rivero, Admiral Horacio 96, 101, 109, 117
Rohwer, Jurgen 71

Schlesinger, James R. 93, 148, 150
sea-launched cruise missiles (SLCM) 3, 122–3
sea lines of communication (SLOC) 10, 12–14, 20, 23–4, 26, 28, 30, 35–6, 39–40, 42, 47–50, 53, 59, 71, 73, 82–3, 89–92, 101, 103, 110–11, 171, 177–81, 183–4, 186–7, 190
Sealift Procurement and National Security Study (SPANS) 142–4, 160, 175
Sealift Readiness Program (SRP) 148–9, 155
Sealift Requirements Study (1966) 139–42
security 14, 48, 63, 100, 127, 190
Senior Civil Emergency Planning Committee (SCEPC) 127, 151; see also PBOS
Sherman, Admiral Forrest P. 8, 24
Sicily 57
Single Integrated Operational Plan (SIOP) 60, 62, 64–5
Sixth Fleet 4, 17, 24–5, 29–30, 32–3, 36, 44, 47, 49, 59–61, 70, 82, 89, 96, 109–10, 179, 181
Somalia 83
Sound Surveillance System (SOSUS) 56, 74, 89
Spain 8, 155
Spitsbergen 12
STEEL SPIKE 139
Standing Naval Force Atlantic (STANAVFORLANT) 57, 104–6, 109, 111, 188
Standing Naval Force Channel (STANAVFORCHAN) 105, 111–12
Strategic Air Command (SAC) 12,

26, 32, 59–60, 62; Emergency
  War Plan (EWP) 59
strategic nuclear deterrence 3, 6, 11,
  14, 24, 30, 42, 46–7, 53–4, 58, 60,
  63, 66–7, 70–3, 88, 90, 93–4, 97,
  100, 113, 118, 125, 130, 187, 189
STRONG EXPRESS 97–8, 121, 167
submarine-launched ballistic missiles
  (SLBM) 3, 55, 61–2, 65, 81, 88,
  97, 114, 122–3, 185–6
Suez Crisis 74
Supreme Allied Commander Atlantic
  (SACLANT) 4, 16–18, 20–30, 32,
  36–7, 40–2, 44, 46, 51–2, 54,
  56–7, 60, 68, 70, 74, 89–90,
  102–6, 109, 112, 117–18, 120,
  122, 158–9, 161–2, 166–8
Supreme Allied Command Europe
  (SACEUR) 4, 14, 16–18, 21,
  23–30, 36–8, 40–4, 47–8, 59,
  61–5, 70, 96, 97, 110, 120–2, 130,
  147, 181
Supreme Headquarters Allied Powers
  Europe (SHAPE) 58, 61, 130
Sweden 39
Syria 83

Tactical Strike Program (TSP) 64–5
Taylor, Maxwell 171
Thatcher government 122
theater-tactical nuclear deterrence 3,
  6, 26, 28, 30, 58, 62, 66, 68, 89,
  95, 97, 102, 122–3, 125, 169, 174,
  185–6, 189
Till, Geoffrey 99
TRADE WIND III 28
Train, Admiral Harry II 120, 181
Tropic of Cancer 122
Trudeau, Pierre Elliott 55
Truman, Harry S. 8, 17, 24
Truman Administration 17
Turkey 29–30, 32, 35, 57, 61, 96,
  109–11, 121, 180; Navy 49, 56

Union of Soviet Socialist Republics
  (USSR) 1, 3, 7–8, 10, 12, 14, 26,
  40, 44, 46, 68–9, 71, 81, 84, 86,
  89, 94, 96–8, 103, 115, 174,
  186–7; Army 14, 91; Naval
  Airforce 10, 178; Navy 1, 10, 12,
  27, 35, 38–41, 47, 55, 71, 74, 76,
  81–92, 100–3, 113–14, 117–18,
  120, 123, 178, 181
United Kingdom 8, 10, 13–14,
  16–17, 21, 36, 46, 48–9, 52–4, 57,
  62–3, 97–8, 104, 111–12, 115,
  117, 119, 121–2, 130, 147, 160,
  163, 172, 187; Royal Air Force
  35; Royal Navy 2, 11, 29, 53–4,
  56, 62, 68, 117
United States 13–14, 17–18, 30, 35,
  46, 49, 52–4, 66, 74, 92, 97, 111,
  119, 120, 122, 126, 130, 133, 135,
  160, 170, 172, 187; Air Force 12,
  15–16, 26–7, 60, 96, 131; Army
  12, 16, 23; Congressional Budget
  Office Study (1976) 93, 95, 178;
  Military Defense Assistance
  Program 49, 117; Navy (USN)
  1–4, 7–9, 11–12, 15–17, chap. 2
  passim, 53, 55, 56–60, 62, 65–7,
  69, 73–4, 81–2, 86, 93, 96, 100,
  105, 116–18, 120–1, 123, 127,
  131, 135, 137, 139, 149, 158–9,
  162–3, 177–9, 183–5, 187–90;
  Weapons Evaluation Group
  (WSEG) 59

Vandenberg, General Hoyt 14
VERITY 16
Vietnam 140

Warsaw Pact 2, 4, 6, 39, 42, 59, 68,
  74–6, 81–3, 86, 91–5, 99, 102,
  110, 123, 145, 170–1, 177, 180,
  187–90
Wartime Use of the US Merchant
  Marine 'Project Walrus' Study
  (1959) 134–5
Watson, Bruce 91
Weeks, Sinclair 134
WELDFAST 35
Western European Union (WEU) 48,
  52
Wilson, Charles 133, 135–6
Wilson, Desmond P. 100
Wilson–Weeks Agreement 134
WINTEX 122, 159

Wolfe, Thomas 86
World Fleet Position Report (1977) 159
Wright, Jerauld 74

Yemen 83

Zakheim, Dov 83, 93
Zumwalt, Elmo 82, 109, 117, 177–8

For Product Safety Concerns and Information please contact our EU representative GPSR@taylorandfrancis.com
Taylor & Francis Verlag GmbH, Kaufingerstraße 24, 80331 München, Germany

www.ingramcontent.com/pod-product-compliance
Lightning Source LLC
Chambersburg PA
CBHW052107300426
44116CB00010B/1560